METHODS IN MOLECULAR BIOLOGY

Series Editor
John M. Walker
School of Life and Medical Sciences
University of Hertfordshire
Hatfield, Hertfordshire, UK

For further volumes:
http://www.springer.com/series/7651

For over 35 years, biological scientists have come to rely on the research protocols and methodologies in the critically acclaimed Methods in Molecular Biology series. The series was the first to introduce the step-by-step protocols approach that has become the standard in all biomedical protocol publishing. Each protocol is provided in readily-reproducible step-by step fashion, opening with an introductory overview, a list of the materials and reagents needed to complete the experiment, and followed by a detailed procedure that is supported with a helpful notes section offering tips and tricks of the trade as well as troubleshooting advice. These hallmark features were introduced by series editor Dr. John Walker and constitute the key ingredient in each and every volume of the Methods in Molecular Biology series. Tested and trusted, comprehensive and reliable, all protocols from the series are indexed in PubMed.

Manuscript Instructions https://drive.google.com/file/d/1K4Az5vpAFVzYadJ3jKx2-QSo7Dqa_SNA/view?usp=sharing

Quantum Dots

Applications in Biology

Fourth Edition

Edited by

Goreti Pereira

Department of Chemistry and CESAM, University of Aveiro, Aveiro, Portugal

Giovannia A.L. Pereira

Department of Fundamental Chemistry, Federal University of Pernambuco, Recife, Brazil

Adriana Fontes

Biophysics and Radiobiology Department, Federal University of Pernambuco, Recife, Brazil

Beate S. Santos

Pharmaceutical Sciences Department, Federal University of Pernambuco, Recife, Brazil

Humana Press

Editors
Goreti Pereira
Department of Chemistry and CESAM
University of Aveiro
Aveiro, Portugal

Giovannia A.L. Pereira
Department of Fundamental Chemistry
Federal University of Pernambuco
Recife, Pernambuco, Brazil

Adriana Fontes
Biophysics and Radiobiology Department
Federal University of Pernambuco
Recife, Pernambuco, Brazil

Beate S. Santos
Pharmaceutical Sciences Department
Federal University of Pernambuco
Recife, Pernambuco, Brazil

ISSN 1064-3745 ISSN 1940-6029 (electronic)
Methods in Molecular Biology
ISBN 978-1-0716-5209-1 ISBN 978-1-0716-5210-7 (eBook)
https://doi.org/10.1007/978-1-0716-5210-7

This Humana imprint is published by the registered company Springer Science+Business Media, LLC, part of Springer Nature.
The registered company address is: 1 New York Plaza, New York, NY 10004, U.S.A.

If disposing of this product, please recycle the paper.

Preface

Quantum dots (QDs) have become essential fluorescent probes across the life sciences due to their remarkable brightness, size-tunable emission, small size, and photostability—features that often surpass those of conventional organic dyes. Since their first applications in biology in the late 1990s, QDs have evolved from proof-of-concept demonstrations to reliable tools for cellular and tissue labeling, biosensing, diagnostic imaging, and theranostics. The 2023 Nobel Prize in Chemistry was awarded to Moungi G. Bawendi, Louis E. Brus, and Alexey Ekimov, recognizing their contributions to QDs, highlighting the fundamental scientific importance and translational impact of these nanocrystals.

Since their first biological applications, numerous strategies have been proposed to enhance (bio)conjugation methods with a wide range of molecular species and nanoparticles. These advancements have led to the development of highly precise and efficient multifunctional systems, suitable for the aforementioned biological applications. Furthermore, with the advent of machine learning techniques, computational approaches have increasingly supported experimental researchers in the ongoing development of more efficient and versatile QDs. Additionally, the synthesis of QDs with novel compositions that reduce the presence of potentially toxic elements has further emphasized the importance of these nanocrystals in advancing improved biological procedures.

In this context, approximately 6 years after the third edition, *Quantum Dots: Applications in Biology, Fourth Edition* presents 16 step-by-step protocols spanning from materials preparation and surface functionalization to bioimaging and sensing. This edition is organized into four sections. The first section covers the preparation of different types of QDs, including binary and ternary compositions, such as PbS, CdTe, Cu_xS, $AgInS_2$, and $CuInS_2$. The second section features three tutorials on using computational tools to optimize optical properties and enable single-QD tracking. The third section compiles chapters dedicated to the functionalization of QDs for bioimaging and cellular labeling, with applications in vitro and in vivo. Finally, the fourth section focuses on the application of QDs for biosensing and biomarker detection.

The editors hope that this volume will serve as a valuable reference and practical resource for students and early-career researchers entering the field, as well as for experienced researchers who already work with QDs. By bringing together well-established methods and emerging trends, the editors expect that this book will serve as a bridge between fundamental research on QDs and impactful applications in the life sciences.

Aveiro, Portugal — Goreti Pereira
Recife, Pernambuco, Brazil — Giovannia A.L. Pereira
Recife, Pernambuco, Brazil — Adriana Fontes
Recife, Pernambuco, Brazil — Beate S. Santos

Contents

Contributors

GABRIELA M. ALBUQUERQUE • *Department of Fundamental Chemistry, Federal University of Pernambuco, Recife, PE, Brazil*

W. RUSS ALGAR • *Department of Chemistry, University of British Columbia, Vancouver, BC, Canada*

JACKIE AUSTIN • *Department of Physics, Creighton University, Omaha, NE, USA*

DUYGU AYDEMIR • *Istanbul University, Institute of Child Health, Department of Pediatric Basic Sciences, Division of Medical Genetics, Istanbul, Turkey*

THOMAS C. BEWAALE • *Department of Chemical Engineering, Northeastern University, Boston, MA, USA*

PAULO EUZÉBIO Cabral Filho • *Department of Biophysics and Radiobiology, Federal University of Pernambuco, Recife, Pernambuco, Brazil*

PAULO H. A. CAMPOS-JUNIOR • *Laboratório de Pesquisas sobre Biologia da Fertilidade em Ambiente Controlado, Departamento de Ciências Naturais, Universidade Federal de São João del Rei, São João del-Rei, MG, Brazil*

MILENE E. M. CASTRO • *Laboratório de Pesquisas sobre Biologia da Fertilidade em Ambiente Controlado, Departamento de Ciências Naturais, Universidade Federal de São João del Rei, São João del-Rei, MG, Brazil*

BRUCE E. COHEN • *The Molecular Foundry, Lawrence Berkeley National Laboratory, Berkeley, CA, USA; Division of Molecular Biophysics & Integrated Bioimaging, Lawrence Berkeley National Laboratory, Berkeley, CA, USA*

SAMANTHA D. COIADO • *Department of Fundamental Chemistry, Federal University of Pernambuco, Recife, PE, Brazil*

LETÍCIA R. C. CUNHA • *Grupo de Pesquisa em Química de Materiais, Departamento de Ciências Naturais, Universidade Federal de São João del-Rei, São João del-Rei, MG, Brazil*

GHINWA H. DARWISH • *Department of Chemistry, University of British Columbia, Vancouver, BC, Canada*

ALLISON M. DENNIS • *Department of Chemical Engineering, Northeastern University, Boston, MA, USA; Department of Bioengineering, Northeastern University, Boston, MA, USA*

KIMAL HONOUR DJAM • *Department of Physics, Creighton University, Omaha, NE, USA*

ANDREW EDET EKPENYONG • *Department of Physics, Creighton University, Omaha, NE, USA*

PORFIRIO ESTRADA-ROJAS • *Universidad Autónoma de Ciudad Juárez, Departamento de Física y Matemáticas, Ciudad Juárez, México*

PAULA FERNÁNDEZ-GÓMEZ • *Instituto Madrileño de Estudios Avanzados en Nanociencia (IMDEA Nanociencia), Madrid, Spain*

ADRIANA FONTES • *Department of Biophysics and Radiobiology, Federal University of Pernambuco, Recife, Pernambuco, Brazil*

Carlos F. G. C. Geraldes • *Department of Life Sciences and Coimbra Chemistry Center (CQC-IMS), Faculty of Science and Technology, University of Coimbra, Coimbra, Portugal*

Cintia Ellen Giarola • *Grupo de Pesquisa em Química de Materiais, Departamento de Ciências Naturais, Universidade Federal de São João del-Rei, São João del-Rei, MG, Brazil*

Oleg Kovtun • *Department of Chemistry, Vanderbilt University, Nashville, TN, USA; Cell Imaging Shared Resource, Vanderbilt University, Nashville, TN, USA*

Bong Han Lee • *Department of Physics, Creighton University, Omaha, NE, USA*

Lin Song Li • *Key Lab for Special Functional Materials of Ministry of Education, and School of Nanoscience and Materials Engineering, Henan University, Kaifeng, China*

Geyse Santos de Lima • *Department of Biophysics and Radiobiology, Federal University of Pernambuco, Recife, Pernambuco, Brazil*

João Victor Araújo de Lima • *Department of Biophysics and Radiobiology, Federal University of Pernambuco, Recife, Pernambuco, Brazil*

Yanbing Lv • *Key Lab for Special Functional Materials of Ministry of Education, and School of Nanoscience and Materials Engineering, Henan University, Kaifeng, China*

Jiao-Jiao Ma • *Department of Radiology, Zhongnan Hospital of Wuhan University, Wuhan, China*

Victor R. Mann • *The Molecular Foundry, Lawrence Berkeley National Laboratory, Berkeley, CA, USA*

Francisco Perazzo Tavares de Melo • *Department of Biophysics and Radiobiology, Federal University of Pernambuco, Recife, Pernambuco, Brazil*

Rebeca M. Melo • *Department of Fundamental Chemistry, Federal University of Pernambuco, Recife, PE, Brazil*

Júlio Batista do Nascimento • *Department of Biophysics and Radiobiology, Federal University of Pernambuco, Recife, Pernambuco, Brazil*

Imelda Olivas-Armendáriz • *Universidad Autónoma de Ciudad Juárez, Departamento de Física y Matemáticas, Ciudad Juárez, México*

Valle Palomo • *Instituto Madrileño de Estudios Avanzados en Nanociencia (IMDEA Nanociencia) Madrid, Spain; Centro de Investigación Biomédica en Red en Enfermedades Neurodegenerativas, (CIBERNED), Instituto de Salud Carlos III, Madrid, Spain*

Amish Patel • *Department of Chemical Engineering, Northeastern University, Boston, MA, USA; Department of Biomedical Engineering, Boston University, Boston, MA, USA*

Cassio C. S. Pedroso • *The Molecular Foundry, Lawrence Berkeley National Laboratory, Berkeley, CA, USA*

Giovannia A. L. Pereira • *Department of Fundamental Chemistry, Federal University of Pernambuco, Recife, PE, Brazil*

Goreti Pereira • *Department of Chemistry and CESAM, University of Aveiro, Aveiro, Portugal*

Ute Resch-Genger • *Federal Institute of Materials Research and Testing (BAM), Division Biophotonics, Berlin, Germany*

Claudia Alejandra Rodríguez-González • *Universidad Autónoma de Ciudad Juárez, Departamento de Física y Matemáticas, Ciudad Juárez, México*

MARCO ANTÓNIO SCHIAVON • *Grupo de Pesquisa em Química de Materiais, Departamento de Ciências Naturais, Universidade Federal de São João del-Rei, São João del-Rei, MG, Brazil*

VALERIA SIGOT • *Institute for Research and Development in Bioengineering and Bioinformatics (IBB), CONICET-UNER, Laboratory of Microscopy (LAMAE). Faculty of Engineering Campus, Oro Verde, Entre Ros, Argentina*

CECÍLIA B. da Silva • *Grupo de Pesquisa em Química de Materiais, Departamento de Ciências Naturais, Universidade Federal de São João del-Rei, São João del-Rei, MG, Brazil*

FERNANDA SUMIKA HOJO DE SOUZA • *Departamento de Computação (DECOM), Instituto de Ciências Exatas e Biológicas, Universidade Federal de Ouro Preto (UFOP)—Campus Universitário Morro do Cruzeiro, Ouro Preto, Minas Gerais, Brazil*

SINDHUJA SURESH • *Department of Physics, Creighton University, Omaha, NE, USA*

ZHI-QUAN TIAN • *College of Chemistry and Molecular Sciences, Wuhan University, Wuhan, China*

AYUSHMAAN TRIPATHI • *Department of Chemical Engineering, Northeastern University, Boston, MA, USA*

NURIYE NURAY ULUSU • *Koc University, School of Medicine, Department of Medical Biochemistry, Sariyer, Istanbul, Turkey; Koç University Research Center for Translational Medicine (KUTTAM), Sariyer, Istanbul, Turkey*

K. DAVID WEGNER • *Federal Institute of Materials Research and Testing (BAM), Division Biophotonics, Berlin, Germany*

RUILI WU • *Key Lab for Special Functional Materials of Ministry of Education, and School of Nanoscience and Materials Engineering, Henan University, Kaifeng, China*

MING-XIA YU • *Department of Clinical Laboratory, Zhongnan Hospital of Wuhan University, Wuhan, China*

Part I

Preparation of QDs and QD-Based Systems

Check for updates

Chapter 1

Synthesis of PbS/CdS/ZnS Core/Shell/Shell Quantum Dots for Multiplexed Shortwave Infrared Imaging

Amish Patel, Thomas C. Bewaale, Ayushmaan Tripathi and Allison M. Dennis

Abstract

This chapter describes a systematic approach for synthesizing high-quality lead sulfide/cadmium sulfide/zinc sulfide (PbS/CdS/ZnS) core/shell/shell quantum dots (QDs) with tunable emission in the shortwave infrared (SWIR) wavelength range (1000–1700 nm). This protocol details the synthesis of three distinct QD formulations with emission peaks at approximately 1150, 1250, and 1515 nm, enabling multiplexed in vivo imaging applications. The synthetic route involves three key steps: (1) size-controlled synthesis of PbS cores through hot injection precipitation, (2) formation of a CdS shell through cation exchange, and (3) deposition of a ZnS outer shell using a low-temperature single-source precursor approach to enhance stability while preserving optical properties. The protocol ends with a postsynthetic surface treatment to enhance quantum yield. These QDs exhibit excellent brightness, narrow emission linewidths, and superior stability compared to conventional SWIR emitters, making them ideal for deep-tissue, multiplexed imaging applications.

Keywords PbS, CdS, ZnS, Core/shell/shell, Multiplex imaging, Size control, Cation exchange, Low-temperature shelling, Nanoparticle stability, Bioimaging

1 Introduction

Fluorescence imaging is a versatile and widely adopted technique for biomedical applications due to its nonionizing nature, cost-effectiveness, and excellent spatial resolution. However, conventional fluorescence imaging using visible and near-infrared (NIR-I, 650–950 nm) wavelengths is limited by tissue autofluorescence, light scattering, and absorption, which collectively restrict imaging depth and signal-to-noise ratio [1–3]. Shortwave infrared (SWIR, 1000–1700 nm) imaging has emerged as a powerful alternative, offering significantly reduced tissue scattering, minimal autofluorescence, and improved tissue penetration depth compared to shorter wavelength modalities [4–7]. These advantages enable high-resolution visualization

Goreti Pereira et al. (eds.), *Quantum Dots*, Methods in Molecular Biology, vol. 3025,
https://doi.org/10.1007/978-1-0716-5210-7_1,

of anatomical structures that were previously challenging to image noninvasively.

The development of bright, stable SWIR emitters is critical to advancing this imaging approach. Among various SWIR-emitting contrast agents—including carbon nanotubes, small molecule dyes, lanthanide-doped nanoparticles, and semiconducting polymer dots—semiconductor quantum dots (QDs) stand out due to their exceptional brightness, photostability, broad excitation spectra, narrow emission profiles, and tunable emission wavelengths [8–12]. Lead chalcogenide-based QDs, particularly PbS, have proven especially valuable for SWIR imaging due to their size-tunable bandgaps that can be adjusted to emit throughout the SWIR window.

Despite their advantages, PbS QDs present considerable challenges for biological applications, including susceptibility to oxidation and potential toxicity from heavy metal content. While PbS/CdS core/shell particles are commonly used for imaging, they can degrade in biological environments, hindering longitudinal studies and potentially releasing toxic ions [12–14]. These limitations can be addressed through a core/shell/shell architecture, where a CdS intermediate shell provides lattice matching and initial passivation while an outer ZnS shell offers enhanced chemical stability, photostability, and reduced toxicity [15, 16]. This PbS/CdS/ZnS structure mirrors other biologically compatible QD compositions, such as CdSe/ZnS and CIS/ZnS, which exhibit superior stability and biocompatibility compared to their non-ZnS-shelled counterparts [17, 18]. However, developing such structures is technically challenging due to the thermal instability of PbS/CdS QDs at the temperatures typically required for ZnS shell growth (>200°C).

In this chapter, we describe a comprehensive protocol for synthesizing PbS/CdS/ZnS core/shell/shell QDs optimized for multiplexed SWIR imaging (Fig. 1). Our approach combines precise size control of PbS cores [19], a controlled cation exchange process for CdS shell formation [20], and a low-temperature ZnS shelling method using zinc diethyldithiocarbamate ($Zn(DDTC)_2$) as a single-source precursor. The resulting QDs exhibit exceptional brightness (quantum yields 10%–20% in aqueous media), narrow emission profiles, and enhanced stability for biological applications. We have detailed the synthesis of three distinct QD formulations with emission peaks at approximately 1150, 1250, and 1515 nm that can be encapsulated with PEGylated phospholipids for excellent colloidal stability and biocompatibility in aqueous environments [21]. These QDs enable a wide range of applications including longitudinal tracking, noninvasive pharmacokinetic studies, biodistribution analysis, and simultaneous visualization of multiple biological targets with exceptional sensitivity and resolution in deep tissue [12].

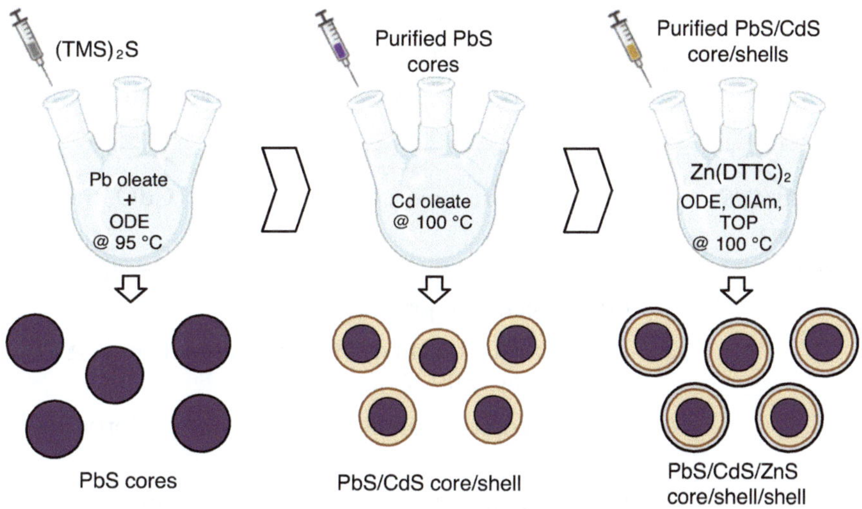

Fig. 1 Schematic overview of PbS/CdS/ZnS core/shell/shell quantum dot synthesis. Schematic representation of the three-step synthesis process for PbS/CdS/ZnS core/shell/shell quantum dots. Step 1: PbS core synthesis by injecting $(TMS)_2S$ into lead oleate at 95°C, followed by purification. Step 2: CdS shell formation through cation exchange by injecting purified PbS cores into cadmium oleate at 100°C, followed by purification. Step 3: ZnS shell growth using $Zn(DDTC)_2$ as a single-source precursor in ODE, OLAm, and TOP at 100°C, resulting in the final core/shell/shell structure. The illustrations show the progressive building of the nanoparticle structure from solid PbS cores (purple) to PbS/CdS core/shell structures (purple/beige) to the complete PbS/CdS/ZnS core/shell/shell particles (purple/beige/gray). Created in BioRender. Dennis A (2025) https://BioRender.com/z435ke2

2 Materials

2.1 Equipment

1. Air-Free Handling System: Schlenk line with vacuum pump (capable of reaching <100 mTorr) and inert gas (argon or nitrogen) supply, vacuum/inert gas manifold with appropriate connections, cold trap, bubbler.
2. Glove Box Setup: Glove box with inert atmosphere (for handling air-sensitive chemicals), antechamber, refrigerator for cold storage, small centrifuge in glovebox, if possible.
3. Characterization Instruments: UV-Vis-NIR spectrophotometer, fluorescence spectrometer with NIR detection capabilities, quartz cuvettes with 1 cm path length.

2.2 Materials for PbS Core Synthesis

1. Lead precursor: Lead(II) oxide (PbO, 99.99%)
2. Sulfur precursor: Bis(trimethylsilyl) sulfide ($(TMS)_2S$, synthesis grade) (*see* **Note 1**)
3. Coordinating ligands and solvents: Oleic acid (OA, technical grade, 90%), 1-octadecene (ODE, technical grade, 90%)

4. Purification solvents: Hexane (anhydrous, ≥99%), ethanol (anhydrous, ≥99.5%)
5. Spectroscopic analysis solvent: Tetrachloroethylene (TCE, ≥99%)

2.3 Materials for CdS Shell Formation via Cation Exchange

1. Cadmium precursor: Cadmium oxide (CdO, trace metal grade, 99.99%)
2. Coordinating ligands and solvents: OA (technical grade, 90%), ODE (technical grade, 90%)
3. Purification solvents: Hexane (anhydrous, ≥99%), ethanol (anhydrous, ≥99.5%)
4. Spectroscopic analysis solvent: Tetrachloroethylene (TCE, ≥99%)
5. Starting material: PbS quantum dot cores (synthesized in Sect. 3.1)

2.4 Materials for ZnS Shell Formation

1. Zinc precursors: For ZnS shell: Zinc diethyldithiocarbamate ($Zn(DDTC)_2$, 97%); For postsynthetic treatment: Zinc acetate (anhydrous, 99.99%)
2. Coordinating ligands and solvents: ODE (technical grade, 90%), oleylamine (OLAm, technical grade, 70%), tri-*n*-octylphosphine (TOP, technical grade, 90%), OA (technical grade, 90%)
3. Purification solvents: Hexane (anhydrous, ≥99%), ethanol (anhydrous, ≥99.5%)
4. Starting material: PbS/CdS core/shell QDs (synthesized in Sect. 3.2)

3 Methods

3.1 PbS Quantum Dot Core Synthesis

This protocol describes the synthesis of three sizes of PbS quantum dots with first excitonic peaks (1S) at approximately 1200, 1400, and 1550 nm. The lead to sulfur ratio in the reaction is constant at 2:1 for all three particles. Particle size and emission wavelength are precisely controlled by adjusting the oleic acid concentration while keeping the injection temperature constant at 95°C. All synthesis steps should be carried out under air-free conditions (*see* **Note 2**). Either nitrogen or argon can be used as the inert gas in this protocol. The reagent amounts used for the synthesis of small, medium, and large PbS cores is summarized in Table 1.

3.1.1 Lead Oleate ($Pb(OA)_2$) Preparation

1. Setup a 100 mL three-neck round-bottom flask with a magnetic stir bar, thermocouple, condenser, and rubber septum. Transfer the flask to a glove box with inert atmosphere.

Table 1
Reagents for PbS core synthesis

	A (small)	B (medium)	C (large)
PbO	0.45 g	0.45 g	0.45 g
Oleic acid	10.1 mL	17.7 mL	23.6 mL
ODE	13 mL	13 mL	13 mL
$(TMS)_2S$ (0.2 mol/L in ODE)	5 mL	5 mL	5 mL
Total reaction volume	28.1 mL	35.7 mL	41.6 mL

2. In the glove box, weigh 450 mg of PbO (2.0 mmol) and add to the flask.
3. Add oleic acid (OA) according to the desired QD size (*see* **Note 3**): For small QDs (~1200 nm 1S peak): 10.1 mL OA (9.0 g, 31.9 mmol); For medium QDs (~1400 nm 1S peak): 17.7 mL OA (15.8 g, 55.9 mmol); For large QDs (~1550 nm 1S peak): 23.6 mL OA (21.1 g, 74.7 mmol).
4. Remove the flask from the glove box and add 13 mL of ODE (*see* **Note 4**).
5. Connect the flask to a Schlenk line with a cold trap and degas the mixture by applying vacuum while stirring. Ensure all connections are tight and pressure is below 100 mTorr.
6. Heat the mixture to 110°C under vacuum and maintain this temperature for at least 20 min to form the lead oleate precursor solution. The solution should become clear and colorless, indicating complete dissolution and formation of lead oleate (*see* **Note 5**).
7. Switch the Schlenk line from vacuum to inert gas (argon or nitrogen) flow. For thorough oxygen removal, cycle between vacuum and inert gas two to three times before finally maintaining the inert gas flow.

3.1.2 Sulfur Precursor Preparation

1. While the lead oleate solution is heating in Step 3.1.1.6, begin preparation of the sulfur precursor (*see* **Note 1**). Remove $(TMS)_2S$ stock from the glove box refrigerator at least 30 min before use to allow it to thaw completely.
2. Measure precisely 0.21 mL (0.18 g, 1.0 mmol) of $(TMS)_2S$ using a syringe.
3. Dilute to 5 mL with ODE in a septum-sealed vial to make a 0.2 M solution. Flush the syringe thoroughly to mix any residual $(TMS)_2S$ from the syringe into the diluting ODE.
4. Draw the solution into a syringe equipped with a 16G needle.

5. Seal the needle by pushing it into a rubber stopper and place the stoppered syringe in a ziplock bag as a secondary containment measure.
6. Place the bagged syringe in the glove box antechamber and purge glovebox for ~10 min.
7. Do not remove the $(TMS)_2S$ syringe from the glovebox antechamber until it is time for the injection.

3.1.3 PbS Nucleation and Growth

1. Reduce the temperature of the lead oleate to 95°C and allow it to stabilize for approximately 10 min to reach thermal equilibrium before injection.
2. Once the lead oleate temperature stabilizes at 95°C, transfer the sulfur syringe from the glovebox antechamber to the fume hood.
3. Remove the heating mantle from the lead oleate solution.
4. Quickly inject the entire sulfur precursor solution into the lead oleate solution.
5. After injection, immediately rinse the empty syringe and needle with hexane and dispose in a sharps container.
6. Allow the mixture to naturally cool to 30°C before proceeding (*see* **Note 6**).

3.1.4 PbS Core Purification

1. Transfer the reaction mixture to the glove box for purification (*see* **Note 7**).
2. Consider setting aside a small volume (100 μL) of the reaction solution for optical measurements (absorbance, photoluminescence, and quantum yield determination). See Sect. 3.1.5 for more detail.
3. Divide the reaction solution among 50 mL centrifuge tubes (two tubes for small QDs, four tubes for medium or large QDs).
4. To each tube, add 5 mL of hexane followed by 15 mL of ethanol (or enough to cause the solution to become turbid) (*see* **Note 8**).
5. Centrifuge the tubes at 2600 rcf for 10 min to precipitate the QDs.
6. Discard the supernatant and resuspend the QD pellets in a total of 10 mL hexane or ODE (*see* **Note 8**).
7. Store the QDs in hexane at 4°C in the glove box refrigerator until needed for the shell growth steps.

3.1.5 PbS Quantum Dot Size and Concentration Determination

Following synthesis, the resulting PbS cores can be characterized using absorption spectroscopy. Use published empirical equations relating the position of the 1S excitonic peak wavelength to the size of the PbS core, an equation for the molar extinction coefficient of the QD at high energy wavelengths (400 nm) or the first excitonic peak, and the Beer-Lambert law to use spectral data to determine the size and concentration of the particles in solution.

1. Take 100 μL of the raw reaction suspension and dilute in 2 mL of TCE. Mix the suspension thoroughly before taking the sample as it is heterogenous. Measure the absorption spectrum using a UV-Vis-NIR spectrophotometer in the range of 400–2000 nm. Note the wavelength of the first excitonic peak (λ_{1S}; Fig. 2) and the absorbance at 400 nm (*see* **Notes 9–11**).
2. Determine the QD size (*d*) using the first excitonic peak position (*see* **Note 12**): Calculate the band gap energy:

$$E_0\,(\text{eV}) = \frac{1240}{\lambda_{1S}}$$

Apply the sizing equation [22]:

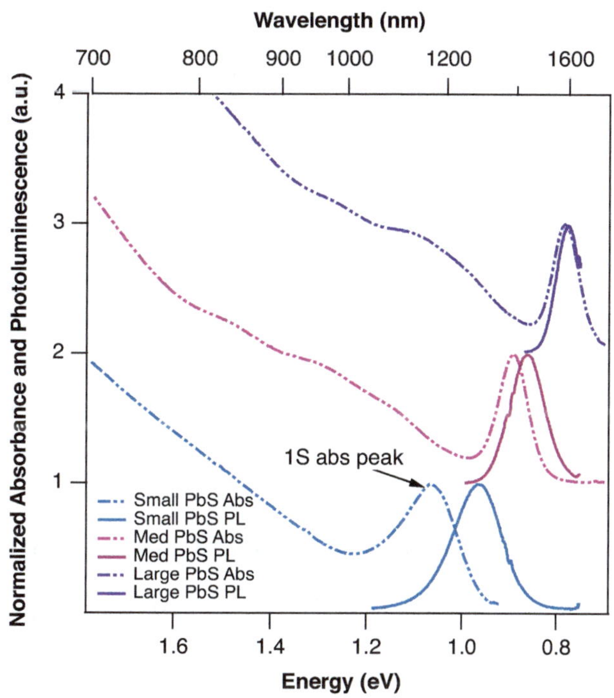

Fig. 2 Optical characterization of PbS quantum dot cores. Absorbance (dashed lines) and photoluminescence emission spectra (solid lines) of small (blue), medium (magenta), and large (purple) PbS quantum dot cores. The first excitonic (1S) absorption peak is indicated by an arrow on the small core spectrum. The 1S absorption peaks appear at approximately 1200, 1400, and 1600 nm for small, medium, and large cores, respectively. Note that the emission spectrum of the large PbS cores is truncated due to detector sensitivity limitations beyond 1600 nm

$$E_0\,(\text{eV}) = 0.41 + \frac{1}{(0.114d + 0.0392d^2)}$$

Solve for d (nm) using a numerical solver or appropriate approximation.

3. Calculate the QD concentration using the absorbance at 400 nm (*see* **Note 13**): Calculate the molar extinction coefficient [23]:

$$\varepsilon_{400} = (0.0233 \pm 0.0001)d^3 \frac{\text{cm}^{-1}}{\mu\text{M}}$$

Apply Beer's Law:

$$\text{Concentration}\,(\mu\text{M}) = \frac{A_{400}}{\varepsilon_{400} \cdot l}$$

where A_{400} is the absorbance at 400 nm and l is the path length in cm (typically 1 cm). Adjust the concentration to account for dilution (*see* **Note 9**):

$$\text{Concentration}_{\text{final}} = \text{Concentration} \times \text{Dilution Factor}$$

4. To calculate the chemical yield based on sulfur conversion [23], determine the appropriate Pb:S ratio (R): use $R = 1.37$ for QDs smaller than 3.7 nm in diameter, or $R = 1.26$ for larger QDs [24]. Calculate the number of S atoms per QD using:

$$S_{\text{QD}} = \frac{4\pi}{3} \times \left(\frac{d}{0.59}\right)^3 \times \frac{1}{(1 + R)}$$

Calculate the total number of QDs generated:

$$N_{\text{QD}} = \text{Concentration} \times \text{Volume} \times N_A,$$

where *Concentration* and *Volume* are the concentration and volume of the QD solution in mol/L and L, respectively, and N_A is Avogadro's number. Calculate the yield:

$$\text{Yield}(\%) = \left(S_{\text{QD}} \times \frac{N_{\text{QD}}}{S_0}\right) \times 100\%,$$

where S_0 is the initial amount of sulfur precursor in moles.

3.2 CdS Shell Formation via Cation Exchange

This section describes the formation of a CdS shell on PbS quantum dot cores via cation exchange. The process involves preparing cadmium oleate as a cation precursor and controlling the reaction temperature to achieve a uniform CdS shell while preserving the core size distribution. The cation exchange reaction is temperature dependent. Higher temperatures accelerate the exchange but may lead to Ostwald ripening and broadening of size distribution. Lower temperatures yield slower exchange but better size uniformity. The 100°C temperature and 5 min reaction time represent an optimal compromise for shell quality and reaction efficiency.

3.2.1 Preparation of Cadmium Oleate Precursor

1. Set up a 250 mL three-neck round-bottom flask with a magnetic stir bar, thermocouple, and condenser.
2. Transfer the flask to a glove box with inert atmosphere.
3. In the glove box, weigh 3.0 g (23.4 mmol) of CdO and add to the flask.
4. Add 22 mL (19.7 g, 63.2 mmol) of OA to the flask, maintaining an OA:Cd molar ratio of approximately 2.7:1 (*see* **Note 14**).
5. Add 95 mL of ODE to the flask to reach a final volume of 117 mL. Seal the flask with a rubber septum.
6. Remove the flask from the glove box and connect to a Schlenk line with a cold trap.
7. Degas the mixture by applying vacuum for 15 min while stirring. Ensure all connections are tight and pressure is at or below 100 mTorr.
8. Switch the Schlenk line from vacuum to inert gas (argon or nitrogen) flow.
9. While under inert gas flow, set the thermocouple to 200°C and maintain heating for 30 min, or until the solution becomes clear and colorless, indicating complete formation of cadmium oleate. Note that the flask temperature may not reach 200°C as a great amount of energy is consumed to convert cadmium oxide.
10. Reduce the temperature to 120°C.
11. At 120°C, switch back to vacuum and degas for 1 h to remove any water produced during the reaction.
12. Allow the solution to cool to room temperature under inert gas flow.
13. Transfer the cadmium oleate solution to the glove box and store in sealed vials until needed for the cation exchange reaction. The final concentration is 0.20 M cadmium oleate in ODE (*see* **Note 15**).

3.2.2 Cation Exchange Reaction

1. Calculate the amount of cadmium oleate needed for the reaction such that there are 6.67 atoms of Cd per atom of Pb in the cation exchange reaction (*see* **Note 16**): Calculate the radius of the PbS cores: $r_{PbS} = d_{PbS}/2$ where d_{PbS} is the diameter determined from the first excitonic peak (Sect. 3.1.5.2). Calculate the volume of a single PbS QD:

$$V_{QD} = \frac{4\pi}{3} r_{PbS}^3 \quad \left(\text{in nm}^3\right)$$

Using the volume of a PbS unit cell (V_{UC}) and the number of Pb atoms per unit cell (0.2134 nm³ and four for the rock salt crystal structure, respectively), calculate the number of Pb atoms per QD (N_{Pb}):

$$N_{Pb} = \frac{\text{Pb atoms per UC} \cdot V_{QD}}{V_{UC}} = \frac{4 \cdot V_{QD}}{0.2134} = 18.74 \cdot V_{QD}$$

Calculate the total moles of Pb atoms (n_{Pb}) in the cation exchange reaction:

$$n_{Pb} = N_{Pb} \cdot n_{QD},$$

where n_{QD} is the moles of PbS QDs (concentration × volume of the PbS core solution) to be used in the cation exchange reaction. Calculate the required moles of Cd for the cation exchange reaction, presuming a 6.67-fold amount of cadmium as lead (*see* **Note 17**):

$$n_{Cd} = 6.67 \cdot n_{Pb},$$

and the volume of cadmium oleate solution needed:

$$V_{Cd(OA)_2} = \frac{n_{Cd}}{[Cd(OA)_2]},$$

where $[Cd(OA)_2]$ is the molarity of the cadmium oleate solution (typically 0.2 M, as prepared in Sect. 3.2.1).

2. To setup the cation exchange reaction, attach a three-neck round-bottom flask with a magnetic stir bar, thermocouple, and condenser to the Schlenk line. Evacuate the flask with vacuum and then refill with argon. Repeat this twice to ensure an air-free environment.

3. Add the appropriate volume of 0.2 mol/L cadmium oleate solution from the glovebox to the flask. Heat the cadmium oleate solution to 100°C (*see* **Note 18**) under inert gas flow (Example reactant amounts included in Table 2).

Table 2
Example reagent amounts for PbS/CdS core/shell generation

	A (small)	B (medium)	C (large)
PbS QDs in hexanes or ODE	1 mL	1 mL	1 mL
PbS core concentration	350 μmol/L	243 μmol/L	46.0 μmol/L
PbS core diameter	5.08 nm	6.00 nm	6.84 nm
0.2 M $Cd(OA)_2$	8.4 mL	10.5 mL	3.8 mL
Total reaction volume	9.4 mL	11.5 mL	4.8 mL

4. After the solution stabilizes at 100°C, inject the PbS QD solution through a septum using a syringe.
5. Allow the reaction to proceed for 5 min at 100°C while stirring.
6. Remove the heating mantle and allow the reaction flask to cool to room temperature gradually. Transfer the flask to the glovebox for purification.
7. In the glove box, transfer the room temperature solution to one or more 50 mL centrifuge tube(s).
8. Add one part (i.e., 1× the reaction volume) hexane and three parts ethanol to precipitate the particles.
9. Centrifuge at 2600 rcf for 10 min.
10. Discard the supernatant and resuspend the PbS/CdS core/shell QD pellet in 5 mL of hexane. Store the PbS/CdS QDs in hexane at 4°C in the glove box refrigerator until needed for ZnS shell growth.
11. Use the core/shell optical properties to determine the CdS shell thickness. Measure the absorption spectrum of the PbS/CdS core/shell QDs and note the wavelength of the first excitonic peak ($\lambda_{1S\text{-}PbS/CdS}$). Calculate the size of the residual PbS core ($d_{PbS\text{-}in\text{-}PbS/CdS}$) using the sizing equation from Sect. 3.1.3. The CdS shell thickness is calculated as:

$$\text{Cd shell thickness} = \frac{d_{PbS} - d_{PbS-in-PbS/CdS}}{2},$$

where d_{PbS} is the diameter of the original PbS cores.

3.3 Depositing a ZnS Shell to Generate PbS/CdS/ZnS core/Shell/Shell Particles

This section describes the formation of a ZnS shell on PbS/CdS core/shell QDs using a low-temperature single-source precursor approach. The use of $Zn(DDTC)_2$ as a precursor allows for ZnS shell growth at temperatures compatible with the thermal stability of PbS/CdS QDs (*see* **Note 19**).

3.3.1 Calculating ZnS Shell Atomic Monolayer Equivalents

Before beginning the ZnS shell synthesis, determine the amount of $Zn(DDTC)_2$ precursor needed based on the desired shell thickness (*see* **Note 20**):

1. The amount of semiconductor shell to be added is usually discussed in terms of atomic monolayers. One atomic monolayer consists of one layer of cations (Zn^{2+}) and one layer of anions (S^{2-}). For ZnS with a cubic crystal lattice parameter a_0 = 0.541 nm, one atomic monolayer is $a_0/2$ = 0.27 nm thick.
2. To calculate the volume of ZnS to be added, first determine the volume of the PbS/CdS core/shell QD: $V_0 = (4\pi/3) \times r_0^3$ and the volume after adding n atomic monolayers (0.27 nm thick each): $V_1 = (4\pi/3) \times (r_0 + 0.27n)^3$. The change in volume is the ZnS shell volume: $V_{ZnS} = V_1 - V_0$.
3. Use the ZnS shell volume to calculate the number of ZnS formula units needed using the volume of one ZnS unit cell (V_{unit} = 0.1583 nm^3) and the number of Zn and S atoms per unit cell (four each). Specifically, the number of ZnS units needed is equal to the volume of ZnS in the shell divided by the volume of a ZnS unit cell times the number of Zn atoms per unit cell: Number of ZnS units needed = $(V_{ZnS} / V_{unit\ cell}) \times 4$
4. Convert the number of ZnS units needed per QD to moles of $Zn(DDTC)_2$ required for the shelling reaction by dividing the number of ZnS units by Avogadro's number and multiplying by the moles of QDs in the reaction (*see* **Note 20**).

3.3.2 ZnS Shell Growth Procedure

1. Inside an argon-filled glove box, weigh $Zn(DDTC)_2$ according to the desired shell thickness into a three-neck round-bottom flask with a magnetic stir bar, thermocouple, and condenser. For example, accounting for an estimated reaction yield of 40%–50%, we would use approximately 47 mg (0.13 mmol of $Zn(DDTC)_2$ to achieve ~2.4 atomic monolayers layers of ZnS on 35 nmol of 5.0 nm diameter QDs (*see* **Notes 21 and 22**).
2. Add 6 mL of ODE, 2 mL of OLAm, and 2 mL of TOP to the flask.
3. Remove the flask from the glove box and connect it to a Schlenk line with a cold trap. Ensure all joints are properly greased and sealed and the stirring speed is set to 500–700 rpm.
4. Apply rigorous air-free technique: Connect the flask to the vacuum line and evacuate for ~1 min while stirring to remove oxygen and moisture. Close the vacuum valve and open the inert gas valve to backfill the flask with argon.
5. Repeat this vacuum/backfill cycle at least three times to ensure a thoroughly oxygen-free environment. After the final backfill, maintain a gentle positive pressure of argon through the

system. Program the heating mantle to heat to 100°C. Start a timer when the temperature begins increasing.

6. When the temperature reaches 80°C (approximately 10 min after the start of heating), inject the PbS/CdS QDs into the flask.
7. Allow the temperature to reach 100°C (approximately 15 min after the start of heating) and maintain this temperature for 45 min.
8. After the reaction is complete, remove the heating mantle and allow the solution to cool to room temperature while maintaining an inert atmosphere.
9. Once cooled to room temperature, either proceed directly to postsynthetic treatment or store the PbS/CdS/ZnS core/shell/shell QDs in the reaction mixture in a refrigerator until further processing (Figs. 3 and 4).

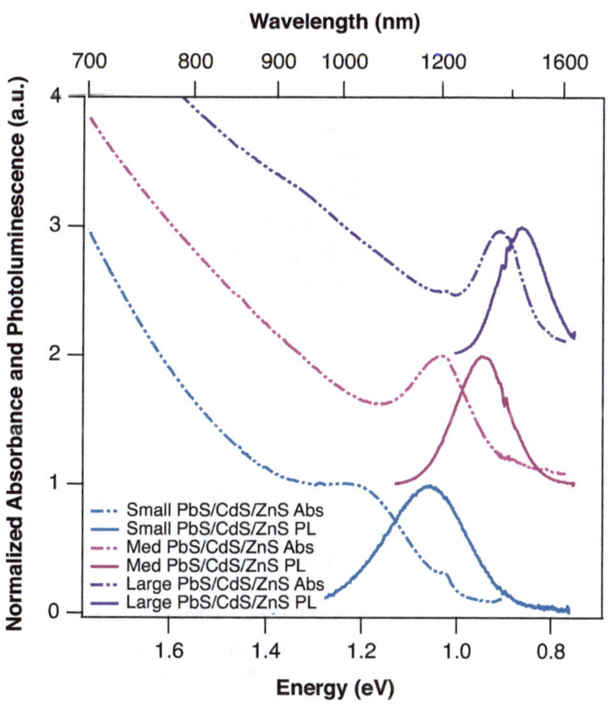

Fig. 3 Evolution of optical properties of core/shell/shell quantum dots. Absorbance (dashed lines) and photoluminescence emission spectra (solid lines) of small (blue), medium (magenta), and large (purple) PbS/CdS/ZnS core/shell/shell quantum dots after complete synthesis. The final particles maintain distinct spectral separation across the SWIR region with emission peaks at approximately 1160, 1240, and 1515 nm, demonstrating that the core/shell/shell QDs maintain the size-dependent optical properties necessary for multiplexed imaging applications

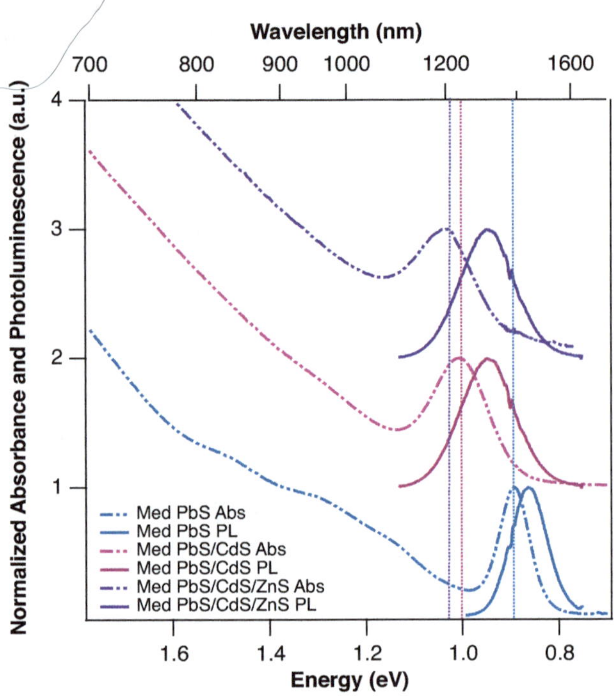

Fig. 4 Spectral evolution during core/shell/shell synthesis. Comparative absorbance and emission spectra showing the progression of medium-sized QDs through each synthesis stage: PbS cores (blue), PbS/CdS core/shell (magenta), and PbS/CdS/ZnS core/shell/shell (purple). Vertical dotted lines mark the 1S peak positions, illustrating a significant blue-shift (~125 nm) during CdS shell formation as the CdS shell encroaches on the PbS core, followed by a minor blue-shift (~25 nm) upon ZnS shell addition. This spectral evolution is indicative of successful shell formation at each synthesis stage while maintaining narrow emission linewidths

3.3.3 Postsynthetic Surface Treatment with Zinc Oleate

A postsynthesis surface treatment serves to further passivate the QD surface and enhance stability and optical properties (Table 3) (*see* **Note 23**):

1. Prepare 0.2 mol/L zinc oleate in TOP (Zn:OA:TOP) for surface treatment of the PbS/CdS/ZnS QDs. Add 3.67 g zinc acetate (20 mmol) and 25.4 mL oleic acid (22.6 g, 80 mmol) to a 250 mL three-neck round-bottom flask with a thermocouple and condenser. Connect the flask to a Schlenk line with a cold trap and degas at 120°C for 1 h to convert zinc acetate to $Zn(OA)_2$. Backfill the flask with argon and add 72 mL of TOP to dilute the solution to 0.2 mol/L. Degas the solution for an additional hour and transfer to an inert gas-filled glove box for long-term storage.

Table 3

Example optical and physical properties of QDs produced using the protocols. Reported PL FWHM and QY for large particles (C) are somewhat higher than reported due to detector cutoff at 1600–1650 nm, indicated with a >sign

Particle	A (small)	B (medium)	C (large)
	PbS cores		
λ_{1S} (nm/eV)	1209/1.026	1364/0.909	1565/0.792
λ_{PL} (nm/eV)	1298/0.955	1420/0.873	1602/0.774
PL FWHM (nm/eV)	150/0.111	171/0.098	>148/>0.067
PbS core diameter (nm)	5.14	5.84	6.84
	PbS/CdS core/shell		
λ_{1S} (nm/eV)	1042/1.190	1239/1.001	1517/0.817
λ_{PL} (nm/eV)	1166/1.063	1313/0.944	1552/0.799
PL FWHM (nm/eV)	196/0.179	185/0.122	>154/>0.081
PL QY (%)	33.79	20.63	>15.67
PbS core radius (nm)	4.44	5.28	6.60
CdS shell thickness (nm)	0.345	0.28	0.13
	PbS/CdS/ZnS core/shell/shell, treated		
λ_{1S} (nm/eV)	1005/1.235	1213/1.022	1477/0.840
λ_{PL} (nm/eV)	1161/1.068	1242/0.998	1515/0.818
PL FWHM (nm)	238/0.217	209/0.169	>154/0.088
PL QY, pretreatment (%)	1.55	2.69	>1.42
PL QY, posttreatment (%)	10.75	14.11	>6.78

2. Purify the QDs for zinc oleate treatment. Precipitate 1–1.5 mL of the PbS/CdS/ZnS QD reaction mixture with an equivalent volume of hexane and excess ethanol to remove excess ligands and unreacted precursors. Centrifuge at 5000×*g* for 10 min and discard the supernatant.
3. Perform the zinc oleate treatment: Add 2 mL of 0.2 M Zn:OA:TOP solution to the QD pellet. Flush the solution briefly with argon. Sonicate the solution in a heated water bath at 60°C for 90 min.
4. Purify the treated QDs by precipitating the QDs twice with equivalent volume of hexane and excess ethanol. Centrifuge at 5000 rcf for 10 min after each precipitation. Discard the supernatant and resuspend the final QD pellet in TCE for optical characterization or hexane for storage and further use.

4 Notes

1. Handle this reagent with caution. It is extremely pungent and toxic. Wear nitrile gloves over the glove box gloves ("double gloving") to protect the permanent glove box gloves from damage. $(TMS)_2S$ requires special handling precautions:
   a. **Health and Safety:** $(TMS)_2S$ is toxic if inhaled or absorbed through skin. Always work in a well-ventilated area or fume hood when handling the syringe outside the glovebox. If any spills occur, clean immediately with a bleach solution to oxidize the sulfide. After injection, immediately rinse the empty syringe and needle with a solvent like hexane and dispose in a designated sharps container.
   b. **Equipment Protection:** $(TMS)_2S$ can damage rubber glove box gloves over time. The double-gloving technique is essential to protect the expensive permanent gloves of the glove box.
   c. **Contamination Prevention:** The odor of $(TMS)_2S$ can persist in a laboratory environment for days if not properly contained. After handling the $(TMS)_2S$, avoid using the glove box for 30–60 min to allow residual odors to dissipate. The smell can contaminate the glove box atmosphere and transfer to other materials if proper precautions are not taken.
   d. **Storage:** Store $(TMS)_2S$ in the glove box refrigerator at 4°C where it solidifies. Remove from refrigeration 30–60 min before use to thaw completely.
   e. **Use of diluted stock solution:** For laboratory efficiency and safety, consider preparing a diluted stock solution: in the glove box, transfer 1 g of $(TMS)_2S$ (approximately 1.18 mL) to 28 mL of ODE in a septum-capped vial to create a 0.2 M solution. This diluted stock can be stored at room temperature in the glove box for up to 3 months and eliminates the need for handling pure $(TMS)_2S$ during each synthesis. When using the 0.2 M $(TMS)_2S$ stock solution in ODE, simply draw 5 mL to obtain the equivalent of 0.21 mL pure $(TMS)_2S$ (1.0 mmol) required for a standard synthesis. The diluted solution has significantly reduced odor, eliminates thawing time, allows for more precise measurements, and improves experimental reproducibility. A 1 g preparation yields enough stock solution for approximately 5–6 standard PbS syntheses.
2. https://www.chemistryviews.org/details/education/4360441/Tips_and_Tricks_for_the_Lab_Air-Sensitive_Techniques_3.html

3. The size of the PbS particles can be increased by increasing the relative amount of oleic acid in the reaction, increasing the injection temperature of the reaction, or both. As described in a prior work [19], the relationship between the oleic acid amount and QD size/1S peak position follows the empirical equation:

$$\lambda_{1S}(\text{nm}) = 935 + m\left(T_{\text{inj}}\right) \cdot g_{\text{OA}},$$

where $\lambda_{1S}(\text{nm})$ is the 1S peak position in nm, T_{inj} is the solution temperature at the time of the sulfur injection, $m(T_{\text{inj}})$ is the slope of the line that describes the relationship between temperature and growth rate/resulting particle size defined as $m(T_{\text{inj}}) = 5.9 + 0.0026T_{\text{inj}}^2$, and g_{OA} is the grams of oleic acid used for the standard reaction comprising 450 mg PbO and 13 mL ODE.

4. ODE can also be added inside the glove box, if preferred.

5. The reaction time is less important than letting the reaction proceed until the solution is clear, indicating that the lead oxide has fully converted to lead oleate. Reactions times in the literature range from 20 min to overnight.

6. This natural cooling is important for size focusing and ensuring monodisperse QDs. Below is a table of times it took for the flask to cool from 95°C to 30°C (Table 4).

7. The "raw" QD solution can be stored in a glass vial at 4°C, where it will solidify into a waxy solid. The PbS cores can be stored under these conditions (air-free, in the reaction solution, in the fridge) for months or even years prior to moving on to the next steps of the reaction.

8. The use of hexane and ethanol specifically is less important than using a less viscous solvent (e.g., hexane, toluene, chloroform) to thin and dilute the oily reaction mixture and approximately three-fold that volume of a polar antisolvent (e.g., ethanol, acetone, methanol/butanol mixture) to aggregate the QDs to enable centrifugation-induced precipitation. The solution will change from a very colored, but clear, solution

Table 4
Reaction cooling following sulfur injection

Temperature (°C)		80	70	60	50	40	30
Time (mm:ss)	Small QD	1:30	3:20	5:45	8:45	15:00	32:00
	Med. QD	2:00	4:15	6:40	10:00	16:20	32:00
	Large QD	2:00	4:50	8:15	10:00	17:00	45:00

to a turbid solution upon addition of sufficient antisolvent to induce aggregation. For postprecipitation resuspension, several solvents are suitable. We most commonly use hexane or ODE. The QDs are more readily soluble in hexane, enabling high concentration stocks, while ODE is a nonvolatile solvent. It can be helpful to have the core QDs in ODE for subsequent reaction steps to avoid hexane boiling and related temperature instability in the short cation exchange reaction.

9. While it is not critical to use the suggested dilution of the QD sample in the absorption measurement, is important to note the specific dilution factor when preparing the sample for optical characterization for use in the concentration determination. The dilution factor equation and its application to our example of 100 μL raw QD solution in 2 mL TCE is as follows:

$$\text{Dilution Factor} = \frac{Vol_{\text{QD}} + Vol_{\text{solvent}}}{Vol_{\text{QD}}} = \frac{0.1\ \text{mL QD} + 2\ \text{mL TCE}}{0.1\ \text{mL QD}} = 21$$

where Vol_{QD} is the volume of the QD sample and Vol_{solvent} is the volume of the diluting solvent, in this case TCE.

10. The reaction solution can be diluted for the optical measurement characterization without purification to determine the size and concentration of the PbS QDs in the reaction mixture. While this is not typical of this synthesis reaction, in cases where unreacted or insoluble reactants cause turbidity that could skew the absorbance measurement, the QD dilution can be centrifuged briefly (e.g., 3000 rcf for 5 min) to pellet any precipitates, removing them from the solution to be analyzed. The supernatant is then transferred to a cuvette for the absorbance measurement.

11. The NIR absorbance features of common solvents can introduce artifacts into the absorption spectrum. We prefer TCE for publication-ready presentation of the absorbance spectrum because of its relative lack of absorbance features in the NIR wavelength range of interest. If the 1S peak of the QD does not overlap with the NIR absorbance features of the solvent, then the measurement can be taken in one's preferred solvent.

12. The equation

$$E_0\ (\text{eV}) = 0.41 + \frac{1}{(0.114d + 0.0392d^2)}$$

uses empirically derived coefficients to fit an equation that adds quantum confinement effects (d term) and Coulombic interactions

(d^2 term) to the bulk bandgap of PbS (0.41 eV) [22, 23]. In practice, it can be easier to use the equation in nm units for λ_{1S} and a direct equation for diameter, as follows:

$$d\ (\text{nm}) = \frac{-0.114 + \sqrt{0.012996 + \frac{0.1568}{\frac{1240}{\lambda_{1S}} - 0.41}}}{0.0784}$$

13. The absorption coefficient at 400 nm (μ_{400}) is size-independent (~$1.7 \times 10^5 \text{cm}^{-1}$) and follows Maxwell-Garnett effective medium theory predictions, confirming bulk-like optical properties at high energies. In contrast, the absorption at the band gap shows strong quantum confinement effects, with the molar extinction coefficient scaling approximately as $\propto d^{1.3}$. For accurate determination of QD concentration, the absorption at 400 nm is preferred over band gap absorption due to its lower sensitivity to size dispersion effects such as 1S peak broadening. While it is the less preferred method, the absorbance intensity at λ_{1S} can be used for concentration determination, as follows:

    a. Calculate the molar extinction coefficient at the band gap:

    $$\varepsilon_{\text{gap}} = 1.8 d^{1.3} \quad \left(\frac{\text{cm}^{-1} \cdot \text{meV}}{\mu\text{M}}\right)$$

    b. Convert to energy-integrated values:

    $$\varepsilon_{\text{gap}} \left(\text{eV} \right) = \frac{\varepsilon_{\text{gap}} \times {E_0}^2 \times 1240}{E_0}$$

    c. Apply Beer's Law using the integrated area under the first absorption peak

14. The OA:Cd molar ratio of 2.7:1 is optimal for forming a stable cadmium oleate complex. Higher ratios can lead to excess ligands in solution, while lower ratios may result in incomplete dissolution of CdO. This can be varied, however. Literature values range from 2.4:1 to 10:1 OA:Cd ratios for different reactions, with 4:1 being a common ratio. The 0.2 M $Cd(OA)_2$ solution is a cadmium-based concentration, so the oleic acid concentration can be increased to change the ratio without changing the 0.2 M cadmium concentration.

15. The cadmium oleate solution can be stored in the glove box for several months without significant degradation. For long-term storage, keep the vials sealed and protected from light.

The precursor recipe can be proportionally scaled down to produce less of the precursor.

16. The molar ratio of Cd:Pb should be approximately 6.67:1 to ensure complete shell formation. The exact volume of cadmium oleate solution needed depends on:
    a. The size of the PbS cores (larger cores require more Cd to form a complete shell).
    b. The number of PbS QDs in the reaction.
    c. The desired shell thickness (typically, 1.5–2 monolayers of CdS).
17. If too large a volume of PbS cores is used, especially in volatile solvent, the temperature control may be inadequate, leading to poorly controlled cation exchange. If this occurs, repeat the reaction paying special attention to temperature control. Consider:
    a. Using more concentrated PbS core solutions.
    b. Removing excess solvent by vacuum prior to injection.
    c. Using the raw PbS core reaction solution directly in a one-pot setup. Instead of adding PbS particles to cadmium oleate, cadmium oleate may be added slowly to heated raw PbS reaction solution to produce similar reaction results, though longer heating (up to 20 min or more) may be needed to produce an equivalent shell thickness.
    d. Redispersing the PbS cores in ODE after cleaning.
18. The cation exchange reaction is temperature dependent. Higher temperatures accelerate the exchange but may lead to Ostwald ripening and broadening of size distribution. Lower temperatures yield slower exchange but better size uniformity. An 80°C temperature and 20-min reaction time represent an alternative set of conditions that yield high shell quality efficiently. Longer times may be used to achieve thicker CdS shells.
19. The low-temperature (100°C) approach using $Zn(DDTC)_2$ as a single-source precursor is critical for preserving the integrity of the PbS/CdS core/shell structure, which can be damaged at the higher temperatures (>200°C) typically required for ZnS shell growth using conventional precursors.
20. The shell thickness can be controlled by adjusting the amount of $Zn(DDTC)_2$ precursor. The atomic monolayer equivalents calculation provides a theoretical guide, but actual shell thickness will vary.
21. For example, to calculate the amount of $Zn(DDTC)_2$ needed to theoretically deposit five atomic monolayers of ZnS

(0.27 nm per monolayer) on PbS/CdS QDs with a diameter of 4.5 nm, subtract the volume of the core/shell from the volume of the desired core/shell/shell, and determine the amount of Zn in the proposed ZnS shell, and calculate the $Zn(DDTC)_2$ as follows:

a. $Vol_1 = Vol_{PbS/CdS} = \frac{4\pi}{3} \times r^3 = \frac{4\pi}{3} \times 2.25^3 = 47.7\ nm^3$

b. $Vol_2 = Vol_{PbS/CdS/ZnS} = \frac{4\pi}{3} \times (2.25 + 5 \times 0.27)^3 = 195.4\ nm^3$

c. $Vol_{ZnS} = Vol_2 - Vol_1 = 195.4\ nm^3 - 47.7\ nm^3 = 147.7\ nm^3$

d. Number of ZnS units = $(147.7\ nm^3 / 0.1588\ nm^3) \times 4 =$ 3719 units

e. Moles of $Zn(DDTC)_2$ = $3719/(6.022 \times 10^{23}) = 6.17 \times 10^{-21}$ mol per QD

f. For 35 nmol of QDs, total $Zn(DDTC)_2$needed = 6.17×10^{-21} mol $\times 35 \times 10^{-9}$ mol $\times 6.022 \times 10^{23} = 0.130$ mmol

g. In our experience, this amount of precursor yields ~2.4 atomic monolayers of ZnS.

22. The thermal decomposition of $Zn(DDTC)_2$ produces byproducts including carbon disulfide, which has a distinctive odor. Ensure adequate ventilation and perform the reaction in a well-functioning fume hood.

23. The postsynthetic zinc oleate treatment is optional but significantly improves quantum yield and long-term stability by passivating remaining surface defects. This step is particularly important for QDs intended for biological applications where stability in aqueous environments is required.

References

1. Saeboe AM, Nikiforov AYu, Toufanian R, Kays JC, Chern M, Casas JP, Han K, Piryatinski A, Jones D, Dennis AM (2021) Extending the near-infrared emission range of indium phosphide quantum dots for multiplexed in vivo imaging. Nano Lett 21:3271–3279. https://doi.org/10.1021/acs.nanolett.1c00600
2. Hong G, Antaris AL, Dai H (2017) Near-infrared fluorophores for biomedical imaging. Nat Biomed Eng 1:1–22. https://doi.org/10.1038/s41551-016-0010
3. Yun SH, Kwok SJJ (2017) Light in diagnosis, therapy and surgery. Nat Biomed Eng 1:1–16. https://doi.org/10.1038/s41551-016-0008
4. Sun Y, Zhong X, Dennis AM (2023) Minimizing near-infrared autofluorescence in preclinical imaging with diet and wavelength selection. J Biomed Opt 28:094805. https://doi.org/10.1117/1.JBO.28.9.094805
5. Welsher K, Sherlock SP, Dai H (2011) Deep-tissue anatomical imaging of mice using carbon nanotube fluorophores in the second near-infrared window. Proc Natl Acad Sci U S A 108:8943–8948. https://doi.org/10.1073/pnas.1014501108
6. Carr JA, Aellen M, Franke D, So PTC, Bruns OT, Bawendi MG (2018) Absorption by water increases fluorescence image contrast of biological tissue in the shortwave infrared. Proc Natl Acad Sci U S A 115:9080–9085. https://doi.org/10.1073/pnas.1803210115
7. Wang F, Zhong Y, Bruns O, Liang Y, Dai H (2024) In vivo NIR-II fluorescence imaging for biology and medicine. Nat Photon 1–13. https://doi.org/10.1038/s41566-024-01391-5
8. Hemmer E, Benayas A, Légaré F, Vetrone F (2016) Exploiting the biological windows: current perspectives on fluorescent bioprobes

emitting above 1000 nm. Nanoscale Horiz 1:168–184. https://doi.org/10.1039/C5NH00073D
9. Bruns OT, Bischof TS, Harris DK, Franke D, Shi Y, Riedemann L, Bartelt A, Jaworski FB, Carr JA, Rowlands CJ, Wilson MWB, Chen O, Wei H, Hwang GW, Montana DM, Coropceanu I, Achorn OB, Kloepper J, Heeren J, So PTC, Fukumura D, Jensen KF, Jain RK, Bawendi MG (2017) Next-generation in vivo optical imaging with short-wave infrared quantum dots. Nat Biomed Eng 1:1–11. https://doi.org/10.1038/s41551-017-0056
10. Zhang M, Yue J, Cui R, Ma Z, Wan H, Wang F, Zhu S, Zhou Y, Kuang Y, Zhong Y, Pang D-W, Dai H (2018) Bright quantum dots emitting at ~1,600 nm in the NIR-IIb window for deep tissue fluorescence imaging. Proc Natl Acad Sci U S A 115:6590–6595. https://doi.org/10.1073/pnas.1806153115
11. Zhang Y, Hong G, Zhang Y, Chen G, Li F, Dai H, Wang Q (2012) Ag_2S quantum dot: a bright and biocompatible fluorescent nanoprobe in the second near-infrared window. ACS Nano 6:3695–3702. https://doi.org/10.1021/nn301218z
12. Zhong X, Patel A, Sun Y, Saeboe AM, Dennis A (2024) Multiplexed shortwave infrared imaging highlights anatomical structures in mice. Angew Chem Int Ed 63:e202410936. https://doi.org/10.1002/anie.202410936
13. Jeong S, Jung Y, Bok S, Ryu Y-M, Lee S, Kim Y-E, Song J, Kim M, Kim S-Y, Ahn G-O, Kim S (2018) Multiplexed in vivo imaging using size-controlled quantum dots in the second near-infrared window. Adv Healthc Mater 7:1800695. https://doi.org/10.1002/adhm.201800695
14. Ma Z, Wang F, Zhong Y, Salazar F, Li J, Zhang M, Ren F, Wu AM, Dai H (2020) Cross-link-functionalized nanoparticles for rapid excretion in nanotheranostic applications. Angew Chem Int Ed 59:20552–20560. https://doi.org/10.1002/anie.202008083
15. Boercker JE, Woodall DL, Cunningham PD, Placencia D, Ellis CT, Stewart MH, Brintlinger TH, Stroud RM, Tischler JG (2018) Synthesis and characterization of PbS/ZnS core/shell nanocrystals. Chem Mater 30:4112–4123. https://doi.org/10.1021/acs.chemmater.8b01421
16. Ren F, del Rosal B, An SY, Yang F, Carrasco E, Benayas A, Oh JK, Jaque D, de la Fuente ÁJ, Vetrone F, Ma D (2017) Development and investigation of ultrastable PbS/CdS/ZnS quantum dots for near-infrared tumor imaging. Part Part Syst Char 34:1600242. https://doi.org/10.1002/ppsc.201600242
17. Kays JC, Saeboe AM, Toufanian R, Kurant DE, Dennis AM (2020) Shell-free copper indium sulfide quantum dots induce toxicity *in* vitro and in vivo. Nano Lett 20:1980–1991. https://doi.org/10.1021/acs.nanolett.9b05259
18. Wang X, Tian J, Yong K-T, Zhu X, Lin MC-M, Jiang W, Li J, Huang Q, Lin G (2016) Immunotoxicity assessment of CdSe/ZnS quantum dots in macrophages, lymphocytes and BALB/c mice. J Nanobiotechnol 14:10. https://doi.org/10.1186/s12951-016-0162-4
19. Zhang J, Crisp RW, Gao J, Kroupa DM, Beard MC, Luther JM (2015) Synthetic conditions for high-accuracy size control of PbS quantum dots. J Phys Chem Lett 6:1830–1833. https://doi.org/10.1021/acs.jpclett.5b00689
20. Ren F, Zhao H, Vetrone F, Ma D (2013) Microwave-assisted cation exchange toward synthesis of near-infrared emitting PbS/CdS core/shell quantum dots with significantly improved quantum yields through a uniform growth path. Nanoscale 5:7800. https://doi.org/10.1039/c3nr02181e
21. Saeboe AM, Kays JC, Dennis AM (2020) Encapsulating quantum dots in lipid–PEG micelles and subsequent copper-free click chemistry bioconjugation. In: Fontes A, Santos BS (eds) Quantum dots: applications in biology. Springer US, New York, NY, pp 95–108
22. Weidman MC, Beck ME, Hoffman RS, Prins F, Tisdale WA (2014) Monodisperse, air-stable PbS nanocrystals via precursor stoichiometry control. ACS Nano 8:6363–6371. https://doi.org/10.1021/nn5018654
23. Moreels I, Lambert K, Smeets D, De Muynck D, Nollet T, Martins JC, Vanhaecke F, Vantomme A, Delerue C, Allan G, Hens Z (2009) Size-dependent optical properties of colloidal PbS quantum dots. ACS Nano 3:3023–3030. https://doi.org/10.1021/nn900863a
24. Choi H, Ko J-H, Kim Y-H, Jeong S (2013) Steric-hindrance-driven shape transition in PbS quantum dots: understanding size-dependent stability. J Am Chem Soc 135:5278–5281. https://doi.org/10.1021/ja400948t

Chapter 2

Preparation of BSA-Cu_xS Quantum Dots via a One-Pot Synthesis Approach Using the Biuret's Reaction

Porfirio Estrada-Rojas, Imelda Olivas-Armendáriz and Claudia Alejandra Rodríguez-González

Abstract

In contemporary chemical synthesis, a significant research focus has emerged on one-pot methodologies. This approach, driven by the desire to enhance efficiency, reduce resource consumption, minimize the generation of toxic by-products, and streamline experimental procedures, facilitates the execution of complex multi-stage reactions within a single reaction vessel. By circumventing the necessity for intermediate purification steps, one-pot synthesis offers a substantial simplification of multistep reactions, synthetic methodologies, and comprehensive syntheses. In the context of biological applications, such methodologies are particularly valuable to produce nanoparticles, such as quantum dots, which have garnered significant attention due to their potential in drug delivery, bioimaging, and biosensing. Consequently, this chapter will delve into the presentation of a one-pot synthetic route for the fabrication of BSA-Cu_2S/CuS quantum dots, using the Biuret reaction.

Keywords Methodology, One-pot synthesis, BSA-Cu_2S/CuS, Quantum dots, Biuret's reaction

1 Introduction

One-pot synthesis, also known as single-step synthesis, is a highly efficient process that enables complex multistep chemical reactions to occur within a single reactor, eliminating the need for intermediate separations (Fig. 1). This approach is cost-effective, environmentally friendly, and streamlines synthesis by avoiding labor-intensive purification steps. Widely used for both organic and organic–inorganic hybrid compounds, one-pot synthesis minimizes resource consumption while maximizing overall process efficiency [1, 2].

In this type of synthesis, the reagent molar ratio and the use of surfactants play a crucial role in promoting synergistic interactions between active sites essential for catalytic reactions. By enabling multiple reaction stages to occur simultaneously, this approach optimizes conditions for each step, leading to a more direct and efficient

Goreti Pereira et al. (eds.), *Quantum Dots*, Methods in Molecular Biology, vol. 3025,
https://doi.org/10.1007/978-1-0716-5210-7_2,

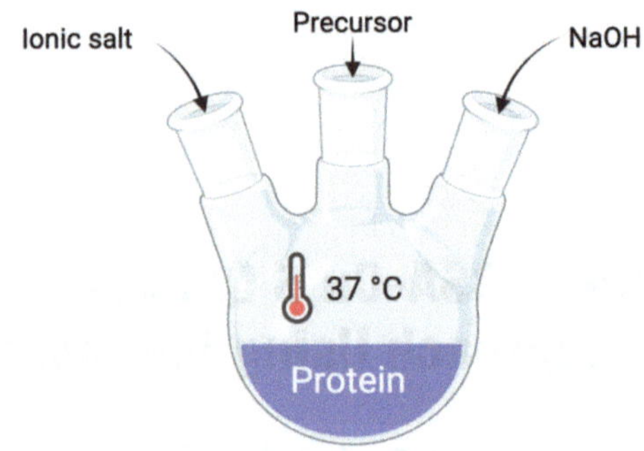

Fig. 1 One-pot synthesis

production of the desired product. Eliminating the need for multiple separation steps further enhances efficiency and reduces overall energy consumption [2].

One-pot synthesis has a wide range of applications across fields such as hydroisomerization, oxygen reduction, CO_2 reduction, and hydrogenation, consistently delivering high activity and conversion rates. However, challenges remain, particularly in enhancing catalyst selectivity and developing efficient methods for separating active sites. Despite these hurdles, its simplicity and effectiveness make one-pot synthesis a valuable and versatile approach in modern chemistry [2, 3].

Illustrative examples of controlled one-pot synthesis for single-atom catalysts (SACs) include nickel embedded in nitrogen-doped carbon nanotubes and nitrogen-doped graphene. These materials are synthesized using citric acid, which promotes the formation of tubular carbon nanotube supports Ni SACs. Notably, even with a lower SAC loading, the increased exposure of active sites enhances electrochemical reduction processes [4].

Another example is the one-pot synthesis of Co-N/C SAC, where urea and glucose serve as carbon and nitrogen sources. This method facilitates the interaction between cobalt and nitrogen atoms, yielding a catalyst with outstanding performance in oxygen reduction and hydrogen evolution reactions. Its efficiency positions it as a promising alternative to commercial catalysts used for these applications [3, 4].

Another noteworthy example is the synthesis of organic, inorganic, and hybrids, such as TiO_2/polyaniline (PAni), achieved through the simultaneous oxidative polymerization of aniline and the hydrolysis of Ti-isopropoxide. This method enabled the formation of PAni nanofibers decorated with TiO_2 nanoparticles, resulting in a high surface area, making them highly suitable for applications like supercapacitors [5].

Beyond these examples, one-pot synthesis has been successfully employed in various advanced materials, including, functionalized copper sulfide quantum dots and magnetic nanoparticles [6–8], jet fuel blends derived from bioderived phenol and cyclopentanol [9], and nanocrystalline composites such as (1−x)PMN-PT/(x)$CoFe_2O_4$ for electronic device applications [10]. These studies highlight the versatility of the one-pot approach, demonstrating its adaptability across diverse fields, from materials for drug delivery systems to catalysis, renewable energy, and advanced material storage.

The one-pot synthesis of heterostructures has also demonstrated its versatility in fields such as Li-S battery materials and microporous conjugated Pd/polymer catalysts. In these cases, single-step synthesis enables precise control of material properties, including the uniform distribution of metal nanoparticles on organic supports, which significantly enhances catalytic efficiency [2, 11].

In summary, one-pot synthesis represents a major advancement in chemistry and materials science by simplifying processes, reducing costs, and improving the sustainability of chemical reactions. Its ability to integrate multiple reaction stages within a single reactor makes it an invaluable tool for developing new functional materials and optimizing industrial processes.

In the other hand, the Biuret reaction is a colorimetric process involving the interaction of specific organic compounds, including biuret, malonamide, oxamide, and related derivatives, with cupric ions (Cu^{2+}) in an alkaline environment. This interaction results in the formation of a colored coordination complex, which is fundamental for both the qualitative and quantitative determination of proteins. The reaction was initially observed by Wiedemann and subsequently characterized in detail by Schiff, who isolated the reaction product of biuret with cupric acetate and potassium hydroxide, establishing its formula as $K_2Cu(biuret)_2$ [12, 13].

Biuret, a compound generated from the thermal decomposition of urea, reacts with copper ions in an alkaline solution, forming a light blue tetradentate coordination complex. This reaction mechanism parallels the interaction between Cu^{2+} and peptide bonds present in peptides and proteins, hence, its significance in protein assays. Polypeptides comprising three or more amino acid residues can form these coordination complexes with cupric ions, leading to a detectable and quantifiable colorimetric change [12].

While the Biuret reaction has historically been regarded as relatively specific for the quantification of proteins and peptides, certain studies document interactions with alternative compounds, including biuret (the namesake of the reaction), amino acids, amino acid amides, dipeptides, and related entities. The characteristics of some of these coordination complexes diverge from those of the canonical copper (II)-biuret complex with proteins, typically manifesting a blue coloration with a maximum absorbance exceeding 600 nm [14].

Within the domain of metallic nanoparticle and quantum dot (QD) synthesis, particularly when functionalized with bovine serum albumin (BSA), the Biuret reaction can be effectively integrated into a one-pot synthetic strategy. In this context, BSA assumes a pivotal role in the stabilization and functionalization of the nanoparticles. The synthesis commences with the dissolution of BSA, a protein-containing peptide bonds, which acts as a stabilizing agent to prevent nanoparticle aggregation. Subsequently, the precursor is introduced (Eq. 1), wherein the metal ion (M^{+}) is indispensable for the formation of the metallic nanoparticles [6].

$$BSA + M^{2+} \rightarrow BSA - M^{2+} \tag{1}$$

Subsequently, sodium hydroxide (NaOH) is introduced to establish an alkaline environment, thereby facilitating the reduction of metal ions (Eq. 2). An ionic salt (I^{-}) is then incorporated, reacting with the metal ions to engender nanoparticle formation (Eq. 3). Beyond its role in nanoparticle stabilization, BSA engages in interactions with metal ions, forming coordination complexes that contribute to the regulation of nanoparticle size and dispersion. Following the nucleation and growth phases of the nanoparticles, the resulting mixture is subjected to lyophilization for subsequent characterization [6].

$$M^{2+} + 2OH^{-} \rightarrow MOH_2 \rightarrow M^{+} \tag{2}$$

$$M^{+} + I^{2} \rightarrow MI \tag{3}$$

Within this framework, the Biuret reaction, which identifies proteins via peptide bond interactions, confirms the effective functionalization of BSA with metal ions. This functionalization, in turn, facilitates the stabilization and regulation of nanoparticle properties, a critical aspect for their application in biomedical domains, including drug delivery systems and oncological diagnostics and therapeutics.

2 Materials

1. NaOH solution: prepare a 1 M solution of NaOH (reagent ≥97%) in distilled water.
2. BSA (purity ≥98%) solution: with a concentration of 25 mg/mL in distilled water (*see* **Notes 1 and 2**).
3. Copper chloride solution: 0.01 mol/L solution of $CuCl_2$ (purity ≥97%) in distilled water.
4. Sodium sulfide nonahydrate solution: prepare a solution of 0.1 mol/L of $Na_2S \cdot 9H_2O$ (purity ≥98%) in 3 mL of distilled water.

3 Methods

3.1 Preparation of BSA-Cu_xS QDs

1. Transfer 5 mL of the prepared BSA solution (25 mg/mL) to a three-necked round-bottom flask and subject it to agitation at 37°C for 10 min (*see* **Note 3**).
2. Following the 10 min agitation period, introduce 5 mL of the $CuCl_2$ solution into the reaction mixture (*see* **Note 4**).
3. After a 2 min period, sufficient for achieving homogeneous mixing of the constituent solutions, introduce 0.5 mL of a 1 mol/L NaOH solution into the reaction mixture (*see* **Notes 5 and 6**).
4. Subsequent to the 10 min reaction period, introduce 0.5 mL of the $Na_2S \cdot 9H_2O$ solution into the reaction mixture. Solution color should in this point change to blue (*see* **Note 7**).
5. At this point, the BSA-Cu_xS QDs start to nucleate (*see* **Note 8**). Solution color turns to a brown-yellowish dark color, which indicates the interaction between free S and Cu ions is occurring.
6. Let the final colloidal dispersion with continuous agitation for 12 h, maintaining a constant temperature of 37°C (*see* **Notes 9 and 10**). After nucleation time, solution's color change to purple. Lower protein concentrations appear to be blue.

3.2 Characterization of BSA-Cu_xS QDs

Physical and chemical characterization is required to verify composition and particle size are as they must be.

1. Fourier transformation infrared spectroscopy (FTIR) can be useful in confirming that the BSA quaternary structure remains intact after synthesis. The amide I and II bands, present at 1650 and 1515 cm^{-1}, respectively, are of particular interest to corroborate BSA presence. Additionally, bands at 1100, 996, and 600 cm^{-1} are typically attributed to the Cu-S bonds, indicating the presence of Cu_xS (*see* **Note 11**). This technique is more effective with freeze-dried samples, as it eliminates interference from water. The interaction between QD's and BSA is suggested by shifts in the bands, particularly those attributed to BSA, as a blue shift may occur due to the vibrational restraint of functional groups, as shown in Fig. 2.
2. To confirm the presence of copper sulfide crystals after synthesis, X-ray diffraction technique (XRD) analysis should be performed. The crystal structures identified by this technique are typically cubic Cu_2S and hexagonal CuS (covellite) (Fig. 3) [15]. XRD must be applied to freeze-dried samples. Peaks from multiple crystal structures can appear simultaneously: therefore, the XRD pattern of this synthesis usually includes

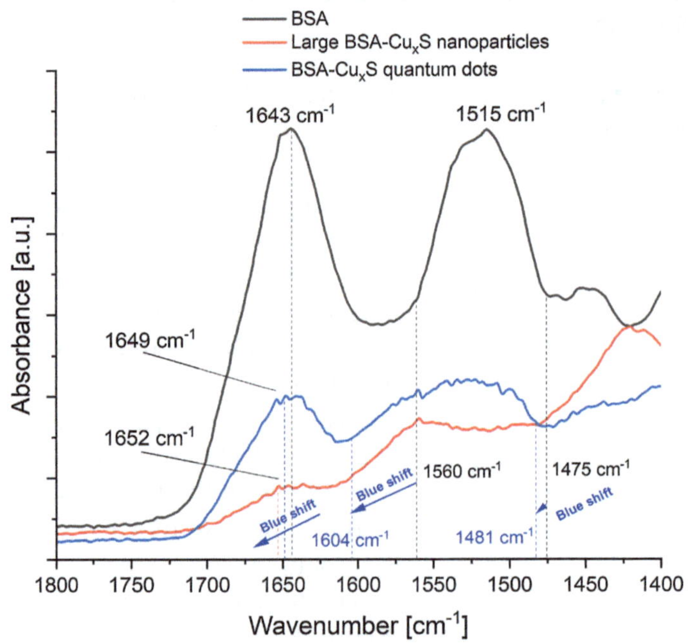

Fig. 2 FTIR spectra of three samples: BSA, a low protein concentration synthesis, and a high protein concentration synthesis. Band displacement is shown (blue shift) attributed to the presence of bonds between Cu_xS crystals and protein

peaks associated with both the cubic and hexagonal phases, as shown in Fig. 4.

3. Dynamic light scattering (DLS) should be used to obtain particle size data. While it provides a good estimate of the size range resulting from this synthesis procedure, the presence of BSA can affect the results by increasing the apparent particle size due to the hydrodynamic radius. This technique should be performed on liquid samples diluted to 0.01%.
4. Transmission electron microscopy (TEM) is a better option for identifying particle size and crystal structures. However, BSA may interfere with the observation of quantum dots, as organic substances tend to degrade under the electron-beam. This technique should be applied to solid samples.
5. Fluorescence spectrophotometry is another method used to identify the presence of quantum dots by measuring the fluorescence of colloidal samples. First, the absorption wavelength associated with excitation should be determined. When this wavelength is applied to a liquid sample, it should exhibit a lower energy fluorescence band.

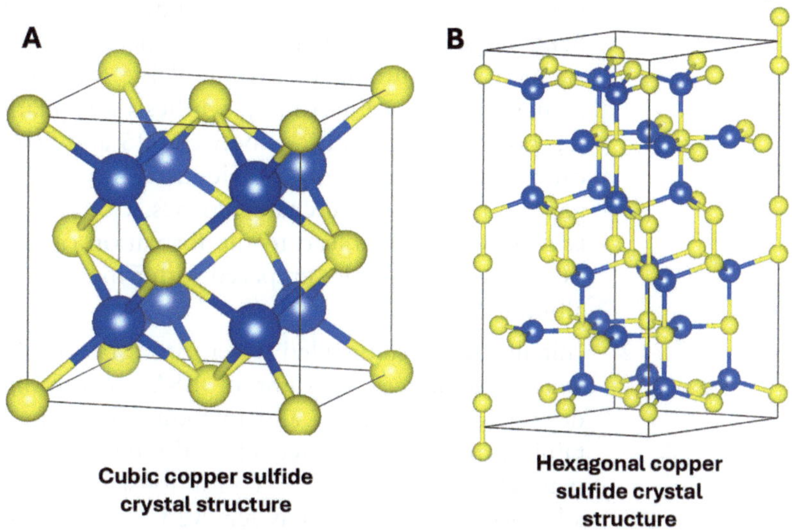

Fig. 3 Cubic Cu_2S (**A**) and hexagonal covellite CuS (**B**) crystal structures obtained during synthesis procedure, according to results obtained by techniques such as X-ray diffraction (XRD) and transmission electron microscopy (TEM)

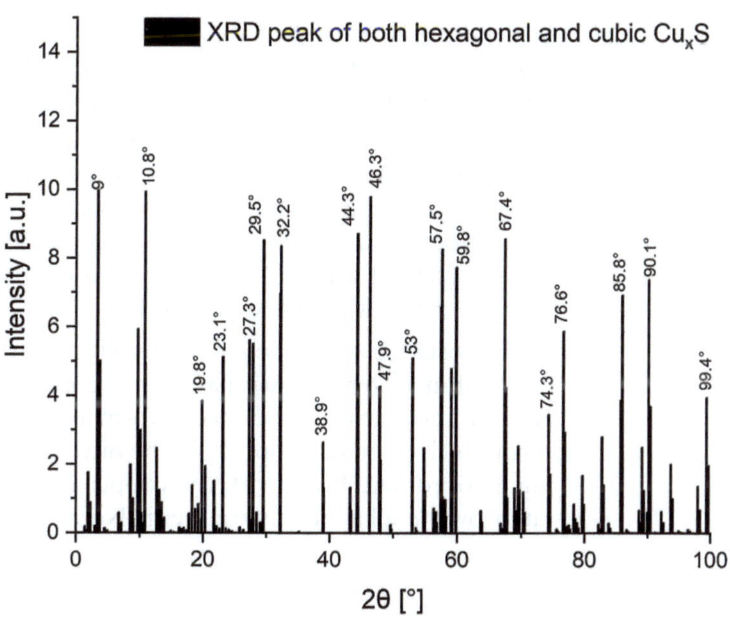

Fig. 4 XRD patterns combined of both hexagonal and cubic Cu_xS structures, as they may appear. Diffraction angles are signaled for those lattice planes which are considered more relevant

4 Notes

1. As a one-pot synthesis methodology, this process requires only a single reaction vessel. Throughout the procedure, a temperature of 37°C and continuous medium agitation are

maintained. The protein concentration serves as a determinant of particle size, with higher concentrations resulting in an increased number of sites available for coordinate bond formation. According to [6], synthesis formulations employing protein concentrations between 25 and 50 mg/mL typically yield quantum dots, as evidenced by particle size, energy gap, and fluorescence characteristics. Conversely, lower protein concentrations lead to the formation of larger particles, as observed in samples prepared with 5 mg/mL bovine serum albumin.

2. Initially, disperse distilled water into a beaker. Subsequently, weigh the precise amount of BSA, corresponding to the desired concentration, and introduce into the beaker containing the distilled water. Place the resulting solutions on a magnetic stirring plate and continuously agitate at 37°C until a homogeneous mixture is achieved.
3. The synthesis mechanism entails the growth of Cu_xS crystals, which are coordinatively bonded through the lone pairs of electrons present on the nitrogen atoms of the peptide bonds within the BSA, as shown in Fig. 5 [15].
4. Upon dissolution of the $CuCl_2$ in the aqueous medium, chlorine (Cl^-) and copper (II) (Cu^{2+}) ions were liberated, facilitating subsequent ionic interactions, as illustrated in Fig. 6 [16].
5. This addition changes the pH from a neutral to an alkaline environment. With the alteration of the electrochemical conditions, coordinate bond formation became thermodynamically favorable. In this context, coordinate bonding requires the presence of a Lewis base to form a coordination adduct with a Lewis acid, as illustrated in Fig. 7 [17].
6. In this instance, the nitrogen atoms within the peptide bonds of BSA function as a Lewis base, while copper (II) ions (Cu^{2+}) act as a Lewis acid. These species form a coordination adduct through electron donation from the nitrogen atom, establishing a coordinate bond with the copper ions, a process facilitated by the alkaline environment. Notably, as this mode of

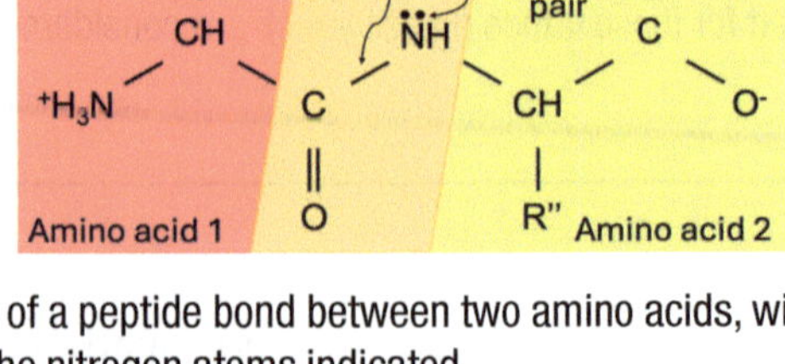

Fig. 5 Example of a peptide bond between two amino acids, with a free electron pair present in the nitrogen atoms indicated

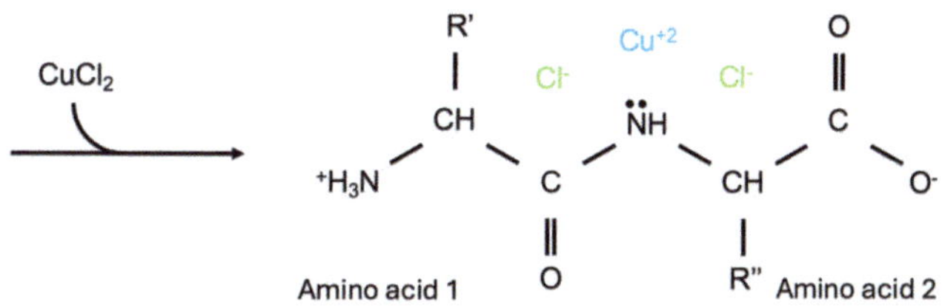

Fig. 6 $CuCl_2$ added solution. As it is observed, Cl^- and Cu^{+2} ions in aqueous solution remain dissociated

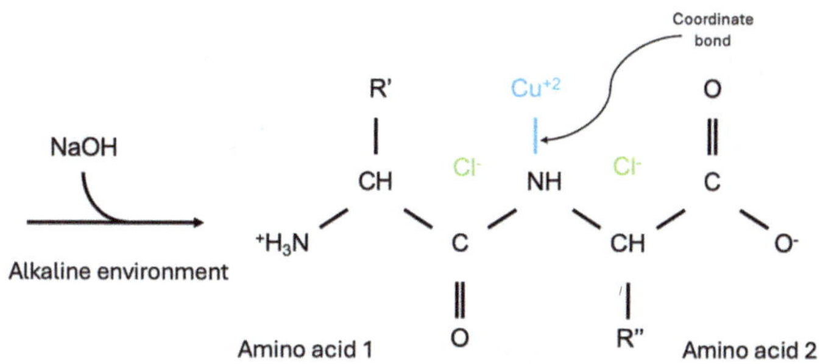

Fig. 7 Coordinate bond formation between a Lewis base (BSA) and a Lewis acid (Cu^{2+}) enhanced by an alkaline environment

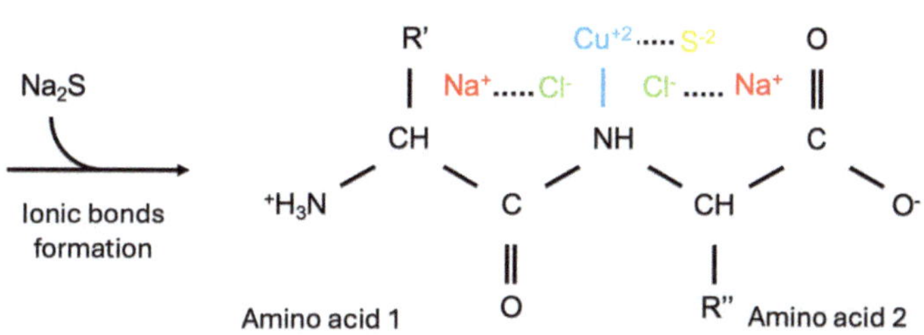

Fig. 8 Suggested ionic interactions after the addition of sodium sulfide

bonding is not ionic, copper retains its capacity to form ionic bonds even when coordinatively bound to BSA [6], as illustrated in Fig. 7.

7. As sodium sulfide nonahydrate is a chalcogenide salt, its constituent ions undergo dissociation within the aqueous environment. This dissociation facilitates the interaction of sulfide ions (S^{2-}) with copper ions (Cu^{2+}) already coordinated to bovine serum albumin (BSA), while sodium ions (Na^+) interact with chloride ions (Cl^-). These ionic interactions are shown in Fig. 8.
8. Ultimately, the formation of the bovine serum albumin–copper sulfide (BSA–Cu$_x$S) conjugated complex occurred,

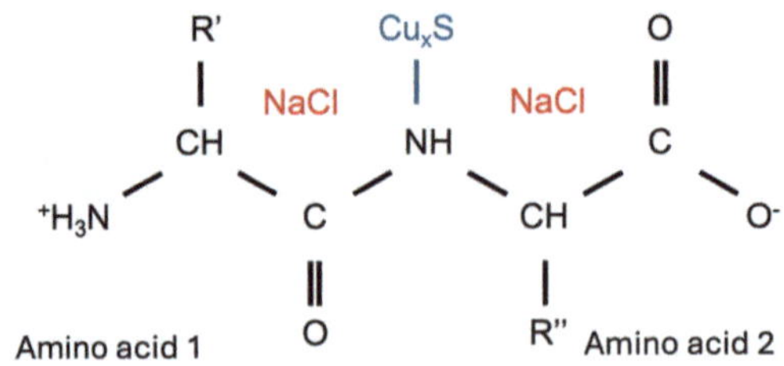

Fig. 9 Formation of copper sulfide coordinate bonded to BSA. As a byproduct, NaCl's formation is suggested

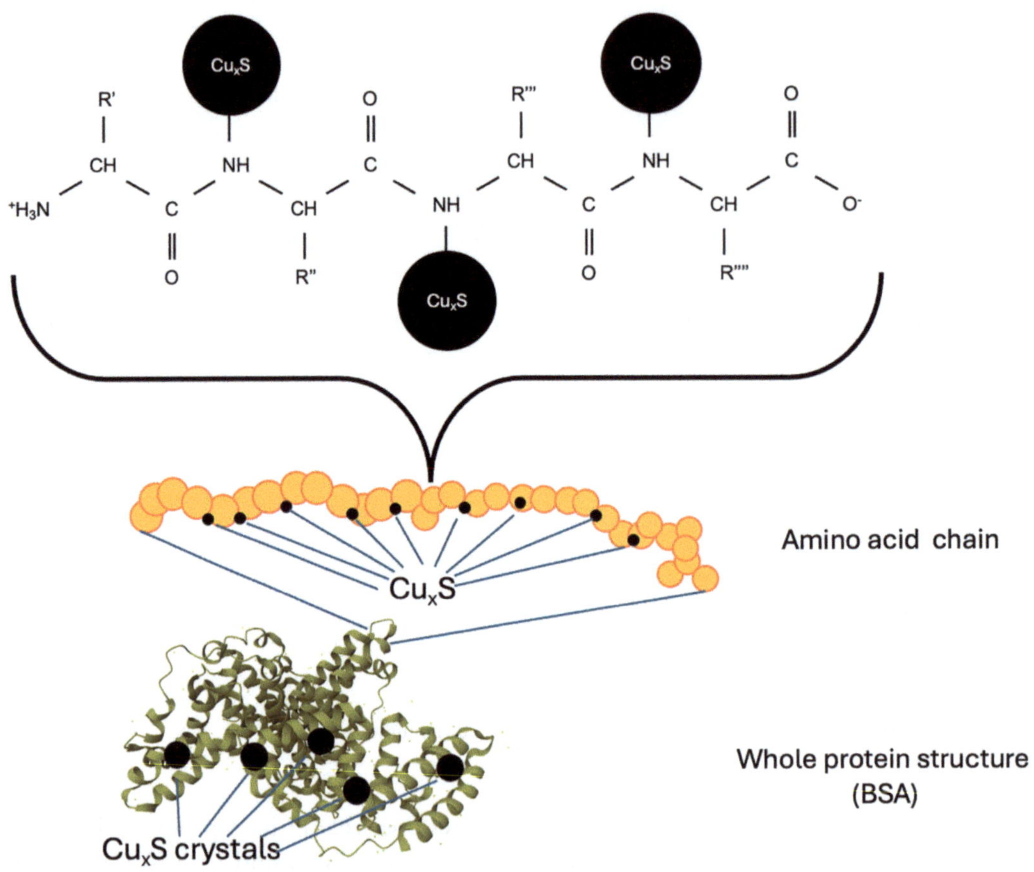

Fig. 10 Cu_xS crystal growing by nucleus formation. BSA protein structure observed is suggested by the Protein Database [20]

with sodium chloride (NaCl) generated as a stoichiometric byproduct of the synthesis reactions, as illustrated in Fig. 9.

9. To facilitate nucleation, precise temporal control is essential [6, 18].
10. The colloidal dispersion was subjected to agitation for 12 h at 37°C, a time required to overcome the activation energy

barrier for nucleation. Given that supersaturated solutions exhibit a propensity for facile nuclei formation, it is anticipated that numerous, small Cu_xS crystal nuclei will develop within solutions containing higher concentrations of BSA. The process of nucleus formation is shown in Fig. 10 [18].

11. It is important to consider this system's chance to be used as a carrier of anticancer drugs. Specifically, it can be applied as any carrier that involves BSA to form ligands with cancer drugs, as it may happen with platinum-based drugs. In order to perform as a drug-carrier, it is necessary to verify BSA and copper sulfide are bonded. This interaction can be suggested to occur through a composition analysis like Fourier's transformation infrared spectroscopy (FTIR) if band shifts are observed [19].

References

1. Samu GF, Janáky C (2017) Tailoring the interfaces in conducting polymer composites by controlled polymerization. In: Hybrid polymer composite materials: structure and chemistry. Elsevier Inc., pp 101–134
2. Mandal K, Ghose S, Mandal M et al (2021) Notes on useful materials and synthesis through various chemical solution techniques. In: Soumen Das, Sandip Dhara (eds) Chemical solution synthesis for materials design and thin film device applications, pp 29–78. https://doi.org/10.1016/B978-0-12-819718-9.00011-X
3. Yang Q, Jiang Y, Zhuo H et al (2023) Recent progress of metal single-atom catalysts for energy applications. Nano Energy 111:108404. https://doi.org/10.1016/J.NANOEN.2023.108404
4. Zhao S, Cheng Y, Veder J-P, Johannessen B et al (2018) One-pot pyrolysis method to fabricate carbon nanotube supported Ni single atom catalysts with ultrahigh loading. ACS Appl Energy Mater 1(10):5286–5297. https://doi.org/10.1021/acsaem.8b00903
5. Bae WJ, Davis AR, Jung J et al (2011) One-pot synthesis of hybrid TiO_2-polyaniline nanoparticles by self-catalyzed hydroamination and oxidative polymerization from TiO_2-methacrylic acid nanoparticles. Chem Commun 47:10710–10712. https://doi.org/10.1039/c1cc13137k
6. Estrada-Rojas P, Hernández-Paz JF, Rodríguez-González CA et al (2025) One-pot synthesis and characterization of a BSA-Cu_2S/CuS quantum dots drug delivery system based on Biuret's reaction. Mater Res Express 12:015406. https://doi.org/10.1088/2053-1591/ada5b0
7. Maleki B, Reiser O, Esmaeilnezhad E, Choi HJ (2019) SO_3H-dendrimer functionalized magnetic nanoparticles (Fe_3O_4@D–NH–$(CH_2)_4$–SO_3H): synthesis, characterization and its application as a novel and heterogeneous catalyst for the one-pot synthesis of polyfunctionalized pyrans and polyhydroquinolines. Polyhedron 162:129–141. https://doi.org/10.1016/j.poly.2019.01.055
8. Ali N, Bilal M, Khan A et al (2021) Fabrication strategies for functionalized nanomaterials. In: Nanomaterials: synthesis, characterization, hazards and safety, pp 55–95. https://doi.org/10.1016/B978-0-12-823823-3.00010-0
9. Nie G, Dai Y, Liu Y et al (2019) High yield one-pot synthesis of high density and low freezing point jet-fuel-ranged blending from bio-derived phenol and cyclopentanol. Chem Eng Sci 207:441–447. https://doi.org/10.1016/j.ces.2019.06.050
10. Fernandez Perdomo CP, A Kiminami RHG, Garcia D (2019) Microwave assisted sintering of nanocristalline PMN-PT/$CoFe_2O_4$ prepared by rapid one pot pechini synthesis: dielectric and magnetoelectric characteristics. Ceram Int 45:7906–7915. https://doi.org/10.1016/j.ceramint.2019.01.101
11. Zeng W, Zhang Y, Zhao X et al (2019) One-pot synthesis of conjugated microporous polymers based on extended molecular graphenes for hydrogen storage. Polymer (Guildf) 174:96–100. https://doi.org/10.1016/j.polymer.2019.04.069
12. Straub J, Weber C, Pusterla N et al (2023) Comparison of refractometer and biuret reaction as measurement methods for serum total protein concentration in warmblood

foals. Pferdeheilkunde 39:5–11. https://doi.org/10.21836/PEM20230101
13. Gornall AG, Bardawill CJ, David MM (1949) Determination of serum proteins by means of the biuret reaction. J Biol Chem 177:751–766. https://doi.org/10.1016/S0021-9258(18)57021-6
14. Hortin GL, Meilinger B (2005) Cross-reactivity of amino acids and other compounds in the biuret reaction: interference with urinary peptide measurements. Clin Chem 51:1411–1419. https://doi.org/10.1373/clinchem.2005.052019
15. Momma K, Izumi F (2011) VESTA 3 for three-dimensional visualization of crystal, volumetric and morphology data. J Appl Crystallogr 44:1272–1276. https://doi.org/10.1107/S0021889811038970
16. Robin AY, Fromm KM, Goesmann H, Bernardinelli G (2003) Structural relationship of two coordination polymers of Cu (i) with the li8and ethanediyl bis (isonicotinate). CrystEngComm 5(71):405-410
17. Arroyo-Currás N, Hall JW, Dick JE et al (2015) An alkaline flow battery based on the coordination chemistry of iron and cobalt. J Electrochem Soc 162:A378–A383. https://doi.org/10.1149/2.0461503jes
18. Brandel C, Ter Horst JH (2015) Measuring induction times and crystal nucleation rates. Faraday Discuss 179:199–214. https://doi.org/10.1039/c4fd00230j
19. Coutts-Lendon CA, Wright NA, Mieso EV, Koenig JL (2003) The use of FT-IR imaging as an analytical tool for the characterization of drug delivery systems. J Control Release 93:223–248. https://doi.org/10.1016/j.jconrel.2003.06.003
20. Bujacz A (2012) Structures of bovine, equine and leporine serum albumin. Acta Crystallogr D Biol Crystallogr 68:1278–1289. https://doi.org/10.1107/S0907444912027047

Chapter 3

Synthesis of Environmentally Friendly Heavy-Metal-Free $AgInS_2$/ZnS Quantum Dots Using Microwave Heating

K. David Wegner and Ute Resch-Genger

Abstract

Luminescent semiconductor nanocrystals, also known as quantum dots (QDs), have had a major impact on applications of optical techniques in the life sciences, such as biosensing and bioimaging. Their unique optical properties, such as broad absorption bands with high molar extinction coefficients, together with their narrow size-dependent photoluminescence (PL) bands, high PL quantum yields (QYs), and a high photostability, are unique advantages for QDs as optical reporters in immunoassays, theranostics, in vitro/in vivo imaging, and flow cytometry. This also laid the foundation for numerous applications of optical multiplexing and barcoding. The most popular QD element compositions are based on heavy-metal elements, such as cadmium or lead, which pose a high toxic risk and are meanwhile restricted in Europe by REACH regulation. Increasing environmental concerns and the growing interest in and need for nanomaterials accessible by sustainable synthesis routes triggered the search for more eco-friendly QDs in the last years. Promising heavy-metal free candidates are ternary $AgInS_2$ QDs, which can be synthesized with high-quality optical properties using aqueous synthesis methods. However, the use of conventional heating mantles or oil baths to heat the reaction mixture to the desired reaction temperature can lead to significant batch to batch variations caused by an inhomogeneous heat diffusion influencing the nucleation and growth of the QDs. A solution to this problem is the use of microwave-assisted heating. This enables a fast and homogeneous heat distribution throughout the entire reaction vessel, resulting in high-quality nanomaterials prepared with high batch-to-batch reproducibility. In this chapter, we will describe the synthesis steps for the preparation of $AgInS_2$/ZnS QDs using microwave-assisted heating. The influence of the stabilizing ligands, variation of precursor concentrations, and the pH of the reaction solution will be detailed in the note section.

Keywords Quantum dots, Microwave-assisted synthesis, Aqueous synthesis, $AgInS_2$, Heavy-metal free, Photoluminescence, Green synthesis

1 Introduction

Since their first application as fluorescent reporters for in vitro imaging, in 1998, semiconductor nanocrystals (also termed quantum dots, QDs) had a large influence on life science applications, such as biosensing and bioimaging [1]. Their unique optical properties like the broad absorption bands with large molar extinction coefficients,

Goreti Pereira et al. (eds.), *Quantum Dots*, Methods in Molecular Biology, vol. 3025,
https://doi.org/10.1007/978-1-0716-5210-7_3,

narrow size-tuneable emission bands, high photoluminescence quantum yields (PL QYs), and high photostability offer great advantages compared to conventional fluorophores, such as fluorescent dyes [2]. The well-established procedures of QD surface modification and bioconjugation laid the foundation to push the boundaries for the sensing of inorganic ions, organic small molecules, and biological macromolecules, cell staining, animal imaging, and therapy [3].

Most biological applications of QDs are based on so-called core/shell QDs. In these heterostructures, the optically active semiconductor core is covered by an epitaxially grown shell made of a second semiconductor material with a larger band gap. This shell has two functions: firstly, it passivates the core surface, and thus, reduces possible defect states that could reduce the optical performance of the QD, and secondly, it increases the chemical stability of the QDs. The latter is important for biological applications where QDs have to withstand enzymes, acidic environments, or reactive oxygen species in biological environments. The most common shelling material for QDs is ZnS [1]. The most popular QD elemental compositions are based on cadmium. Even when shelling these QDs with multiple layers of ZnS, the risk of ion leaching from the core to the environment cannot be excluded. This raised great concerns against their use in biological applications [4].

Among the different types of heavy-metal free QDs, ternary $AgInS_2$/ZnS (AIS) QDs are particularly popular as they can be prepared in aqueous solution with high-quality PL properties. Whereas the PL properties of binary QDs are dependent on the width of their band gap, for ternary QDs, the PL properties are ascribed to defect states in the band gap structure and by the element composition [5–9]. In general, they can be prepared with PL QY over 50% and a PL emission ranging from the visible until the near infrared (NIR) region from about 400 nm to 1000 nm. In contrast to binary QDs their PL bands are broad with a full at half-maximum (FWHM) of more than 100 nm. Despite their long PL lifetimes of several hundred nanoseconds, ideal for time-gated emission, the PL of AIS QDs is not affected by oxygen [10]. The PL mechanisms responsible for these optical properties are still debated. Possible origins are the radiative recombination of donor-acceptor pairs, self-trapped exciton models, and/or a recombination of a localized hole with a conduction band electron [10, 11].

The synthesis of ternary QDs is mostly performed by hot injection or heating up methods in high-boiling organic solvents. This results in hydrophobic QDs, which need to be rendered hydrophilic before they can be used in biological applications. Among different approaches for QD surface modification, the surface ligand exchange is still the most popular method. Thereby, the hydrophobic ligands from the initial QD synthesis, such as oleylamine or 1-dodecanthiol, are substituted for hydrophilic ligands, such as glutathione (GSH), 3-mercaptopropionic acid (MPA), or mercaptoacetic acid. However, this ligand displacement can result in the formation of surface defects diminishing QD optical properties, and hence, QD performance [3]. In this respect, the direct synthesis of hydrophilic QDs is advantageous. In addition, it also minimizes the use of toxic organic hazardous solvents and chemicals, which is increasingly important in terms of the green synthesis approach [12, 13]. Green synthesis describes the use of eco-friendly reaction pathways involving mild reaction conditions, less toxic chemicals, and lower waste amounts. Following this idea, researcher started to prepare nanomaterials using aqueous wet-synthesis approaches [7, 14, 15], hydrothermal methods [16, 17], and microwave-assisted (MW) syntheses [18–21].

In particular, the use of MW radiation is advantageous due to the very fast and homogeneous increase of the reaction temperature forming a stable thermal gradient in the reaction mixture. This enables a more uniform QD synthesis and increases the reproducibility of the synthesis. MW-assisted approaches exist for binary QDs but also for ternary QDs reaching PL QYs of up to 50% [22].

The synthesis of ternary QDs is very challenging due to different reactivities of the used metal ions because they have different hard and soft Lewis acid and base (HSAB) character [23]. For AIS QDs, it has to be considered that In(III) is a so-called hard Lewis acid while Ag(I) is a soft one, which has consequences for the initial precursor concentration used in QD synthesis. This requires a nonstoichiometric ratio of the metal ions, using more indium than silver [23, 24]. As mentioned before, the element composition is important for the PL properties, and thus, the precursor concentrations need to be precisely controlled to obtain reproducible results. In contrast to binary QDs, also surface shelling, i.e., the synthesis of a second semiconductor shell is challenging for AIS QDs. Most of the time, the zinc treatment results in a cation exchange between zinc ions and metal ions forming the QD core. This leads to an alloy formation and a rather quaternary nanomaterial [8, 25, 26]. Whereas in organic solvents, the "active" precursor concentration is largely not affected by the environment, in aqueous solution many more parameters need to be considered. For example, the stabilizing ligands, such as MPA, used during the synthesis can form complexes with silver, indium, and zinc [27, 28], with the complex formation largely depending on the ligand concentration [29]. Therefore, the so-called feeding thiol-In ratio is also important for balancing the thiol binding strength to the different cations present in the reaction mixture [24]. Additionally, the pH can influence the thiol protonation state and the formation of metal hydroxides, which can quickly form precipitates in the reaction solution. Due to the hard base character of hydroxides, the effective concentration of free indium largely depends on the pH [23].

Taking all these influences into account renders the optimization of such syntheses very tedious and time-consuming. Soares et al. used a design of experiment (DoE) approach to rationally prepare AIS/ZnS QDs and precisely tuning the PL features [10]. Thereby, mathematical models were developed based on experimental data, which enabled to predict the optical properties of the AIS QDs such as the PL QY when changing MPA-In ratio, Ag-In ratio, and pH. This significantly reduced the overall number of experiments necessary to optimize the AIS QD synthesis [10].

The protocol for the synthesis of $AgInS_2$/ZnS QDs presented here is based on our experiences gained over many years of preparing ternary QDs and published in this reference [10], among synthesizing many other QDs [22, 26, 30–32]. We could show that the sequential order, precursor concentration, and choice of the surface ligand all have a significant influence on the optical properties of the synthesized QDs. Nevertheless, more systematic studies are necessary to better understand the formation of AIS QDs and to identify the chemical nature of the monomers, formed complexes, and possible even clusters during the reaction.

2 Materials

All solutions need to be prepared with ultrapure water, which should be made by purifying deionized water to obtain a conductivity of 18 MΩ·cm at 25°C, and analytical grade reagents if not otherwise stated. Chemical stock solutions can be prepared at room temperature in the presence of oxygen. There is no need to work in a Glovebox. Stock chemicals should be stored in their designated safety cabinets, whereas the prepared chemical stock solutions should be freshly prepared and used within the same day. Due to the absence of heavy-metal elements, the prepared aqueous stock solutions and the waste produced during the purification process can be discarded as alkaline aqueous waste. Concentrated bases and acids should be diluted and neutralized before discarding them to the aqueous waste. In the following, we will list the necessary equipment for the synthesis and the preparation of the necessary stock solutions, which are then used in the Methods part for the actual synthesis. The molarity of the stock solution is given in the brackets.

2.1 Equipment

1. Single-Mode microwave with 35 mL vessel or comparable (*see* **Note 1**).
2. Centrifuge for 15 and 50 mL Falcon tubes, max. spin-speed for 50 mL Falcon tubes 15,600 ×*g*.
3. Multiposition magnetic stirring plate.
4. Magnetic stirring plate with stand clamp.

2.2 Stock Solutions

1. Nitric acid solution (0.1 mol/L): Add dropwise 105 µL of nitric acid (65%) to 15 mL ultrapure water and stir the solution for 2 min to ensure a homogeneous mixture (*see* **Note 2**).
2. Ammonia hydroxide solution (1.0 mol/L): Add dropwise 526.3 µL ammonia hydroxide solution (NH_4OH, 25%) to 6473.7 µL ultrapure water, and stir the solution for 2 min (*see* **Note 3**).
3. Silver solution (0.1 mol/L): Weigh in 17.0 mg of silver nitrate (99.8% purity), add 1 mL of ultrapure water, and stir until the silver nitrate is dissolved (*see* **Note 4**).
4. MPA solution (1.0 mol/L): Add 2.28 mL ultrapure water to 218 µL of 3-mercaptopropionic acid (MPA, 99% purity) and stir at least for 2 min (*see* **Note 5**).
5. Indium solution (1.0 mol/L): Add 110.6 mg indium (III) chloride (98% purity) to 500 µL of nitric acid solution (prepared in Step 1) and stir until the indium (III) chloride is dissolved (*see* **Note 6**).
6. Sulfide solution (1.0 mol/L): Add 360.3 mg of sodium sulfide nonahydrate ($Na_2S{\cdot}9H_2O$; 98% purity) to 1.5 mL of ultrapure water and stir until the sodium sulfide is dissolved (*see* **Note 7**).
7. Zinc solution (1.0 mol/L): Add 297 mg of zinc nitrate hexahydrate (99% purity) to 1 mL ultrapure water and stir until the zinc nitrate is dissolved (*see* **Note 8**).

3 Methods

The synthesis of AIS QDs can be divided into three steps. In the first step, the actual AIS core QD is synthesized, followed by the shelling procedure, and finally the purification of QDs.

3.1 Core Synthesis

1. Calibrate the pH meter with pH buffers ranging from pH 4 to pH 11. Make sure that the electrode fits into the microwave vessel (*see* **Note 9**).
2. Place a 35 mL microwave vessel with a size-appropriate magnetic stirring bar over a stirring plate using a stand clamp (*see* **Note 10**).
3. Add 25 mL of ultrapure water into the microwave vessel and start the stirring of the stirring plate.
4. Add 400 µL of the silver solution (0.04 mmol) prepared in Sect. 2.2.
5. Add 1000 µL of MPA solution (1.0 mmol) prepared in Sect. 2.2 (*see* **Note 11**).

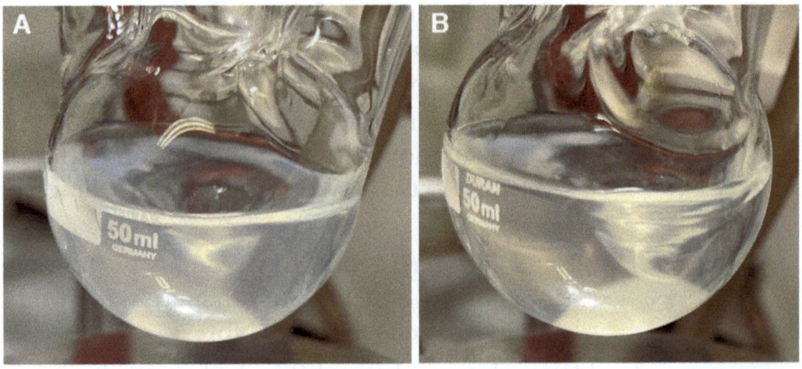

Fig. 1 Pictures of the reaction solution before (**A**) and after (**B**) addition of ammonia hydroxide to the mixture of silver solution and MPA solution. The solution needs to be transparent before proceeding to the next step

6. Add slowly 1000 μL of ammonia hydroxide solution prepared in Sect. 2.2 and stir until the solution becomes clear/transparent again (Fig. 1).
7. Add dropwise 200 μL of indium solution (0.2 mmol) prepared in Sect. 2.2. Wait between the additions so that the white precipitates formed can either be dissolved or at least homogeneously mixed in the reaction mixture (*see* **Note 12**).
8. Adjust the pH of the reaction solution to 8.5 by dropwise addition of the prepared ammonia solution in Sect. 2.2.
9. Add swiftly 500 μL of sulfide solution (0.5 mmol) prepared in Sect. 2.2. At this point, the solution changes its color from transparent to yellow/orange.
10. Put the cap on the microwave vessel and place it into the microwave and heat the solution to 100°C using an initial power output of 200 W. After reaching 100°C, hold the temperature for 50 min before cooling to at least 60°C and proceeding with the shelling procedure.

3.2 Shell Synthesis

1. Place the microwave vessel with the unpurified reaction solution from Sect. 3.1 over a stirring plate using a stand clamp and remove the cap.
2. Add dropwise 200 μL of MPA solution (0.2 mmol) prepared in Sect. 2.2 under stirring.
3. Add 200 μL of ammonia solution prepared in Sect. 2.2 and stir for 2 min.
4. Add 200 μL of zinc solution (0.2 mmol) prepared in Sect. 2.2 and stir for 2 min.
5. Put the cap on the microwave vessel and place it into the microwave and heat the solution to 100°C using an initial power output of 200 W. After reaching 100°C, hold the temperature for 30 min before cooling down to room temperature.

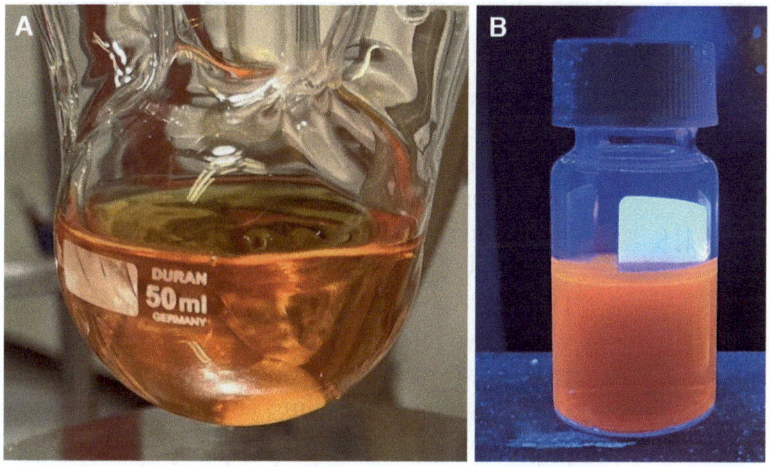

Fig. 2 (**A**) Picture of the QD solution after the shelling procedure before purification. (**B**) Picture of the QD solution after purification under a UV-lamp with an excitation wavelength of 365 nm

3.3 Purification

1. Take the reaction solution and split the solution equally into three 50 mL Falcon tubes. Each tube filled with approximately 10 mL of reaction solution.
2. Add 20 mL of ethanol (96% purity) to each of the Falcon tubes and shake for 10 s using a vortex shaker (medium speed).
3. Weigh the Falcon tubes and make sure that all three have the same weight. In case one of them is too low in weight, add more ethanol to the Falcon tube.
4. Place Falcon tubes into a centrifuge and centrifuge them for 5 min at 10,000 × *g* (*see* **Note 13**).
5. Discard the supernatant by decanting without disturbing the precipitate, place the tube upside down on a piece of paper towel and tap lightly to remove the remaining supernatant from the tube (*see* **Note 14**).
6. Place the Falcon tubes into a holder and add 3 mL of ultrapure water into each tube.
7. Shake the solution gently so that the precipitated QDs can redisperse.
8. If larger precipitates remain, sonicate the tubs for 1 min in an ultrasonication bath.
9. Combine the contents of all three tubes in one glass vial and store the QDs in the dark and in the refrigerator until further use. The transparent solution should be orange/red and have a strong emission when placed below a UV-lamp (Fig. 2).

4 Notes

1. Microwave systems are available from different manufacturers, important is that these are single-mode microwave systems and not household microwaves designed to heat up liquids or food in private environments. For this protocol, the CEM microwave system Discover SP with a 35 mL attenuator was used.
2. Due to the work with concentrated acids, always add the water first into the glass vial and add afterward the concentrated acid to avoid strong exothermic reactions. This could result in violently boiling and uncontrolled splashing of the solution. The used pipette tip should be neutralized in water before throwing it into the chemical contaminated trash bin, as acid residues may remain on the tip.
3. Due to the heavy odor of concentrated ammonia solution, please prepare the solution in a fume hood or using local exhaust ventilation.
4. Silver is highly sensitive to light; the solution should be, thus, covered during stirring with aluminium foil until use. When the stock chemical has changed its color from white to brown/black, silver nitrate is decomposed into silver oxide and nitric acid.
5. MPA is sensitive to oxygen and can undergo oxidation, leading to the formation of disulfides. Therefore, the solution cannot be stored for a long time, and also, the stock chemical needs to be replaced after multiple uses.
6. Complete dissolution of indium (III) chloride is only possible in the presence of nitric acid. Otherwise, indium hydroxide will form and precipitate as dark gray powder. If indium will not completely dissolve, more nitric acid needs to be added. Indium (III) chloride is hygroscopic, and the stock chemical should be stored under inert gas. It is sufficient to close the cap of the chemical under an argon flow, which displaces the air within the container.
7. Sodium sulfide nonahydrate is highly hygroscopic and needs to be stored in the fridge. When taking the chemical from the fridge, wait until the bottle warmed up to room temperature to avoid the condensation of atmospheric water inside the container. Similar to indium (III) chloride, store the chemical under inert gas by flushing the bottle with argon before closing the cap.
8. Similar to the before mentioned chemicals, zinc nitrate hexahydrate is also highly hygroscopic and should be stored under inert gas. It is sufficient to close the cap of the chemical under an argon flow, which displaces the air within the container.

9. The buffer solutions used to calibrate the pH meter should be taken from the buffer bottle. Never put used pH buffer back into the bottle. Make sure that the slope of the calibration line is at least above 93%. If the slope is below this value, the pH electrode should be cleaned according to the manufacturer's protocol.
10. Depending on the microwave model used, the size of the reaction vessel may vary. The protocol presented here has been optimized for the Discover SP from CEM. If a different manufacturer is used, make sure that the reaction vessel has a capacity of at least 30–35 mL.
11. When adding MPA to the silver solution a white cloudy precipitation can be observed. This is due to the chelation of MPA with the silver ions, which are soluble in presence of ammonia hydroxide.
12. It is important to very slowly drop the indium solution into the reaction mixture, so that the formed indium hydroxide can directly dissolve. At the end of the addition, the reaction solution can look slightly turbid but become transparent again when the pH is adjusted to 8.5 (Fig. 3).
13. After centrifugation, there should be a nearly clear/yellowish supernatant and an orange/reddish precipitate. If the supernatant is still very orange or red, either add more ethanol or centrifuge for longer.
14. At this step, it is important to remove the remaining supernatant as the presence of ethanol after QD redispersion, in the next step, could result in a limited colloidal stability of the AIS QDs. However, avoid allowing the precipitate to become too dry, as this would hinder the redispersion of the AIS QDs in the next step.

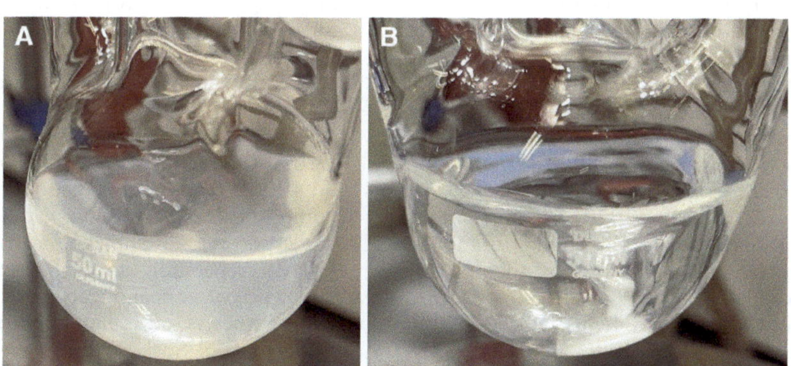

Fig. 3 Pictures of the reaction solution before (**A**) and after (**B**) adjusting the pH to 8.5. The solution needs to be transparent before proceeding to the next step

Acknowledgments

U.R.G. and K.D.W. acknowledge financial support by the DFG-ANR grant SIREN (DFG RE1203/38-1).

References

1. Wegner KD, Resch-Genger U (2024) The 2023 nobel prize in chemistry: quantum dots. Anal Bioanal Chem 416(14):3283–3293. https://doi.org/10.1007/s00216-024-05225-9
2. Resch-Genger U et al (2008) Quantum dots versus organic dyes as fluorescent labels. Nat Methods 5(9):763–775. https://doi.org/10.1038/nmeth.1248
3. Wegner KD, Hildebrandt N (2015) Quantum dots: bright and versatile *in vitro* and *in vivo* fluorescence imaging biosensors. Chem Soc Rev 44(14):4792–4834. https://doi.org/10.1039/c4cs00532e
4. Hong SN et al (2023) Monitoring leaching of Cd^{2+} from cadmium-based quantum dots by an Cd aptamer fluorescence sensor. Biosens Bioelectron 220:8. https://doi.org/10.1016/j.bios.2022.114880
5. Sandroni M, Wegner KD, Aldakov D, Reiss P (2017) Prospects of chalcopyrite-type nanocrystals for energy applications. ACS Energy Lett 2(5):1076–1088. https://doi.org/10.1021/acsenergylett.7b00003
6. Berends AC et al (2019) Optoelectronic properties of ternary I-III-VI_2 semiconductor nanocrystals: bright prospects with elusive origins. J Phys Chem Lett 10(7):1600–1616. https://doi.org/10.1021/acs.jpclett.8b03653
7. Raevskaya A et al (2017) A fine size selection of brightly luminescent water-soluble Ag-In-S and Ag-In-S/ZnS quantum dots. J Phys Chem C 121(16):9032–9042. https://doi.org/10.1021/acs.jpcc.7b00849
8. Song JLQ et al (2016) Bandgap and structure engineering via cation exchange: from binary Ag_2S to ternary $AgInS_2$, quaternary AgZnInS alloy and AgZnInS/ZnS core/shell fluorescent nanocrystals for bioimaging. ACS Appl Mater Interfaces 8(37):24826–24836. https://doi.org/10.1021/acsami.6b07768
9. Raievska O et al (2020) Composition-dependent optical band bowing, vibrational, and photochemical behavior of aqueous glutathione-capped (Cu, Ag)-In-S quantum dots. J Phys Chem C 124(35):19375–19388. https://doi.org/10.1021/acs.jpcc.0c05453
10. Soares JX et al (2020) Rationally designed synthesis of bright $AgInS_2$/ZnS quantum dots with emission control. Nano Res 13(9):2438–2450. https://doi.org/10.1007/s12274-020-2876-8
11. Stroyuk O et al (2018) Origin and dynamics of highly efficient broadband photoluminescence of aqueous glutathione-capped size-selected Ag-In-S quantum dots. J Phys Chem C 122(25):13648–13658. https://doi.org/10.1021/acs.jpcc.8b00106
12. Wang C, Gao X, Su XG (2010) *In vitro* and *in vivo* imaging with quantum dots. Anal Bioanal Chem 397(4):1397–1415. https://doi.org/10.1007/s00216-010-3481-6
13. Zhang HL et al (2010) Special method to prepare quantum dot probes with reduced cytotoxicity and increased optical property. J Biomed Opt 15(1):8. https://doi.org/10.1117/1.3291999
14. Stroyuk O et al (2018) Origin of the broadband photoluminescence of pristine and Cu+/Ag+ -doped ultrasmall CdS and CdSe/CdS quantum dots. J Phys Chem C 122(18):10267–10277. https://doi.org/10.1021/acs.jpcc.8b02337
15. Zhu CN et al (2017) Aqueous synthesis of highly fluorescent and stable Cu-In-S/ZnS core/shell nanocrystals for cell imaging. RSC Adv 7(80):51001–51007. https://doi.org/10.1039/c7ra10215a
16. Song JLQ et al (2015) Facile synthesis of water-soluble Zn-doped $AgIn_5S_8$/ZnS core/shell fluorescent nanocrystals and their biological application. Inorg Chem 54(4):1627–1633. https://doi.org/10.1021/ic502600u
17. Yang YL et al (2019) Facile surface engineering of Ag-In-Zn-S quantum dot photocatalysts by mixed-ligand passivation with improved charge carrier lifetime. Catal Lett 149(7):1800–1812. https://doi.org/10.1007/s10562-019-02718-6
18. Mousavi-Kamazani M, Salavati-Niasari M (2014) A simple microwave approach for synthesis and characterization of Ag_2S-$AgInS_2$ nanocomposites. Compos Pt B-Eng 56:490–496. https://doi.org/10.1016/j.compositesb.2013.08.066
19. Tadjarodi A, Cheshmekhavar AH, Imani M (2012) Preparation of $AgInS_2$ nanoparticles by a facile microwave heating technique; study of effective parameters, optical and photovoltaic characteristics. Appl Surf Sci 263:449–456. https://doi.org/10.1016/j.apsusc.2012.09.080
20. Xiong WW, Yang GH, Wu XC, Zhu JJ (2013) Microwave-assisted synthesis of highly luminescent $AgInS_2$/ZnS nanocrystals for dynamic

intracellular Cu(II) detection. J Mat Chem B 1(33):4160–4165. https://doi.org/10.1039/c3tb20638f
21. Mir IA et al (2018) Bandgap tunable AgInS based quantum dots for high contrast cell imaging with enhanced photodynamic and antifungal applications. Sci Rep 8:12. https://doi.org/10.1038/s41598-018-27246-y
22. Dhamo L et al (2022) Assessing the influence of microwave-assisted synthesis parameters and stabilizing ligands on the optical properties of AIS/ZnS quantum dots. Sci Rep 12(1):11. https://doi.org/10.1038/s41598-022-25498-3
23. Jing LH et al (2016) Aqueous based semiconductor nanocrystals. Chem Rev 116(18):10623–10730. https://doi.org/10.1021/acs.chemrev.6b00041
24. Regulacio MD et al (2013) Aqueous synthesis of highly luminescent $AgInS_2$-ZnS quantum dots and their biological applications. Nanoscale 5(6):2322–2327. https://doi.org/10.1039/c3nr34159c
25. Mao BD et al (2013) Study of the partial Ag-to-Zn cation exchange in $AgInS_2$/ZnS nanocrystals. J Phys Chem C 117(1):648–656. https://doi.org/10.1021/jp309202g
26. Miropoltsev M et al (2022) Influence of hydrophilic thiol ligands of varying denticity on the luminescence properties and colloidal stability of quaternary semiconductor nanocrystals. J Phys Chem C 126(47):20101-20113. https://doi.org/10.1021/acs.jpcc.2c05342
27. Fernando Q, Freiser H (1958) Chelating properties of beta-mercaptopropionic acid. J Am Chem Soc 80(18):4928–4931. https://doi.org/10.1021/ja01551a042
28. Sarin R, Munshi KN (1972) Physicochemical Investigation on complexes of indium (III) with mercapto-substituted, hydroxy-substituted, and amino-substituted propionic acid. Aust J Chem 25(5):929–939. https://doi.org/10.1071/CH9720929
29. Kojima N, Sugiura Y, Tanaka H (1978) Polynuclear and mononuclear complex-formation between indium(III) and sulfhydryl-containing bidentate ligands. Chem Pharm Bull 26(2):579–584
30. Wegner KD et al (2019) Gallium—a versatile element for tuning the photoluminescence properties of InP quantum dots. Chem Commun 55(11):1663–1666. https://doi.org/10.1039/c8cc09740b
31. Sandroni M et al (2018) Cadmium-free $CuInS_2$/ZnS quantum dots as efficient and robust photosensitizers in combination with a molecular catalyst for visible light-driven H_2 production in water. Energy Environ Sci 11(7):1752–1761. https://doi.org/10.1039/c8ee00120k
32. Wegner KD et al (2019) Influence of the core/shell structure of indium phosphide based quantum dots on their photostability and cytotoxicity. Front Chem 7:12. https://doi.org/10.3389/fchem.2019.00466

Chapter 4

Preparation of $CuInS_2$ Quantum Dots with Reduced Toxicity for Biological Applications

Letícia R. C. Cunha, Cecília B. da Silva, Milene E. M. Castro, Paulo H. A. Campos-Junior and Marco Antônio Schiavon

Abstract

This chapter describes the synthesis and characterization of copper indium sulfide $CuInS_2$ (CIS) quantum dots (QDs), as well as their application in cellular labeling and cytotoxicity assessment. Specifically, the synthesis process was optimized to obtain QDs with desirable optical properties. The characterization is mainly based on X-ray diffraction (XRD) and spectroscopy techniques, such as absorption, emission, and time-resolved photoluminescence measurements. Additionally, cytotoxicity tests were performed to ensure its suitability for biological applications. Finally, cell labeling experiments were conducted to demonstrate the potential of $CuInS_2$ QDs as efficient fluorescent probes for bioimaging.

Keywords Quantum dots, CIS, Fluorescence, Cytotoxicity, Live-cell imaging

1 Introduction

Quantum dots (QDs) have attracted the interest of the scientific community, since their discovery in 1981 [1]. QDs exhibit broad absorption bands with high extinction coefficients, narrow and symmetric emission bands, excellent photostability, and high photoluminescence quantum yields (PLQYs). Their most remarkable feature is the ability to tune optical properties based on size, as the nanometric scale induces the spatial confinement of charge carriers, leading to the quantum confinement effect. With decreasing size, the confinement effect intensifies, resulting in a larger bandgap and a corresponding blue shift in both the absorption onset and emission bands. This size-dependent behavior allows precise adjustment of the photoluminescence properties within a specific range of the electromagnetic spectrum [2].

The adjustable light emission of the QDs enables precise tracking within biological systems, making them interesting for biological applications, such as bioimaging [3, 4], biosensing [5], drug

Goreti Pereira et al. (eds.), *Quantum Dots*, Methods in Molecular Biology, vol. 3025,
https://doi.org/10.1007/978-1-0716-5210-7_4,

administration [6], and SARS-CoV-2 diagnosis [7]. QDs can also be modified with therapeutic agents, antibodies, or peptides to specifically target cellular pathways, thereby, improving treatment effectiveness while reducing side effects. By integrating diagnostic and therapeutic functions into a single platform, QDs represent a highly versatile and impactful approach [8].

In the initial phases of QDs synthesis, researchers have focused primarily on exploring binary QDs and their derivative systems. However, with advancements in science and technology, binary QDs are composed of environmentally hazardous and biologically toxic elements such as Cd, As, and Pb [9]. These challenges have prompted nanomedicine researchers to explore new systems that offer improved biocompatibility while maintaining stable, bright, and tunable emission properties. As a result, Cd-free and Pb-free QDs, such as silver, silicon, and germanium-based nanoparticles, have been investigated. The most promising alternatives are copper indium sulfide ($CuInS_2$, CIS) QDs.

$CuInS_2$, a I-III-VI type semiconductor QD (Fig. 1), exhibits lower cytotoxicity, enhanced biocompatibility, and greater environmental stability than traditional heavy metal QDs, such as Cd, Hg, and Pb [4]. In addition, it has a suitable bandgap (Eg = 1.5 eV), high light absorption coefficients ($\alpha > 105\ cm^{-1}$), and excellent photostability in the visible spectrum [10]. Most importantly, for biomedical applications, the emission of CIS QDs covers the red-to-near-infrared (NIR) spectral range, aligning with the so-called biological transparency window—wavelengths that penetrate biological tissues most effectively. Additionally, their photoluminescence exhibits long lifetimes (hundreds of nanoseconds), which is crucial for enhancing bioimaging techniques by improving the signal-to-noise ratio [11].

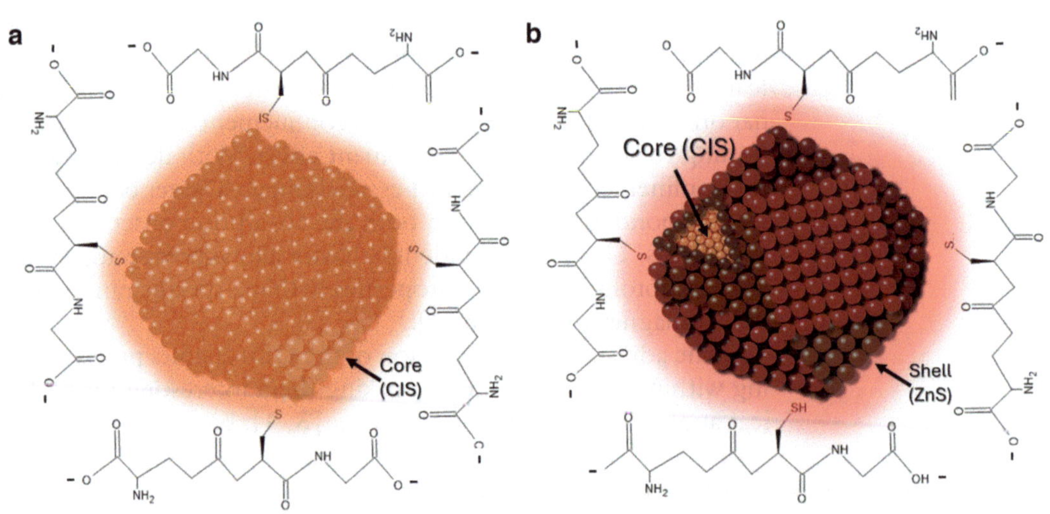

Fig. 1 Schematic representation of **(a)** CIS/ZnS **(b)** CIS/ZnS QDs containing surface ligand (Glutathione—GSH)

To improve the properties of nanomaterials, surface engineering has been used, optimizing passivation by adding a shell, forming a core/shell structure (Fig. 1) [12]. As a result, holes and electrons are tightly confined in the nucleus region, which means that this type of material presents high PLQYs, greater stability, and luminescence intensity. High-density polymers, peptides, antibodies, and oligonucleotides can be added to the core/shell structure to increase hydrophilicity and improve stability in biological systems.

As reported in the literature, the primary approach for synthesizing ternary QDs is organic-phase synthesis. However, this method is both complex and expensive and often requires costly or toxic organic reagents. In addition, most procedures require high temperatures and inert gases, requiring stringent reaction conditions. Ternary QDs are typically water-insoluble, necessitating further surface modification to facilitate their dispersion in aqueous solutions. One common strategy involves replacing hydrophobic ligands on the QD surface with hydrophilic molecules, thereby, altering their surface properties and enabling their transfer to the water phase [13].

However, employing these methods increases the complexity of the QD preparation and introduces certain drawbacks to ternary QDs. These effects include a weakened fluorescence emission spectrum, lower-than-expected water dispersibility, and reduced colloidal stability in aqueous solutions. Therefore, the direct synthesis of ternary QDs using an aqueous-phase method is crucial for overcoming these limitations. Water substitutes costly and toxic organic reagents, and the lower reaction temperature simplifies the experimental process. Additionally, the ternary QDs synthesized by this method are water-dispersible, thereby, eliminating the need for complex posttreatment procedures.

Below we detail the protocols for the synthesis and purification of CIS and CIS/ZnS QDs using glutathione (GSH) as a surface ligand. We also describe the cytotoxicity analysis procedure using the MTT assay. The resulting CIS/ZnS QDs will be tested using Chinese hamster ovary (CHO-K1) cell marking.

2 Materials

2.1 Synthesis and Characterization of $CuInS_2$/ZnS

1. Copper solution: Prepare a 1.0 mol/L stock solution of $CuCl_2.2H_2O$ by dissolving 0.1705 g of the precursor in 1 mL of ultrapure water.
2. Indium solution: Prepare a 1.0 mol/L stock solution of $InCl_3.4H_2O$ by dissolving 0.2932 g of the precursor in 1 mL of ultrapure water.
3. GSH solution: Prepare a 0.25 mol/L stock solution of GSH, by dissolving 0.075 g of the precursor in 1 mL of ultrapure water (*see* **Note 1**).

4. Sodium sulfide solution: Prepare a 1.0 mol/L stock solution of $Na_2S.9H_2O$ by dissolving 0.4801 g of the precursor in 2 mL of ultrapure water.
5. Zinc solution: Prepare a 0.5 mol/L stock solution of $ZnCl_2$ by dissolving 0.1363 g of the precursor in 1 mL of ultrapure water.

2.2 MTT Assay

1. Chinese hamster ovary (CHO-K1) cells (ATCC—CC1-61).
2. Dulbecco's Modified Eagle Medium (DMEM).
3. Fetal bovine serum.
4. 1% antibiotic (100 IU/mL penicillin, 100 μg/mL streptomycin).
5. MTT Solution: 3-[4,5-dimethylthiazol-2-yl]-2,5-diphenyl tetrazolium bromide (MTT) 5 mg/mL in phosphate buffered saline (PBS).
6. Solution of sodium dodecyl sulfate (SDS) (10% HCl).
7. Prepare a stock concentration of QDs by dissolving 1 mg of the CIS/ZnS QDs in 1 mL of DMEM.

2.3 Live Cell In Vitro Imaging

1. Prepare a stock concentration of QDs by dissolving 1 mg of the CIS/ZnS QDs in 1 mL of DMEM culture medium.

3 Methods

3.1 Growth and Purification of $CuInS_2$-ZnS Quantum Dots

To prepare the QDs note the following steps. Firstly, grow the core $CuInS_2$ (CIS), followed by overcoating with a ZnS layer.

3.1.1 CIS Core

1. Add 10 mL of ultrapure water to a glass vial. Add 20 μL of the copper chloride solution, 100 μL of the indium chloride solution, and 800 μL of GSH solution (*see* **Note 2**).
2. Adjust the pH of the mixture to ten using NH_4OH 27%.
3. Finally, add 200 μL of Na_2S solution to the mixture.
4. Add a magnetic stirring bar to the flask, and seal with thread-sealed tape.
5. The mixture was kept at 100°C with constant stirring for 2 h in a water bath (Fig. 2) (*see* **Note 3**).
6. At the end of the synthesis, the solution acquires a darker color.

3.1.2 ZnS-Overcoating

After the synthesis of the CIS core, a thin layer of ZnS (shell) is grown on the nanocrystals. The use of a shell to cover nanocrystals is

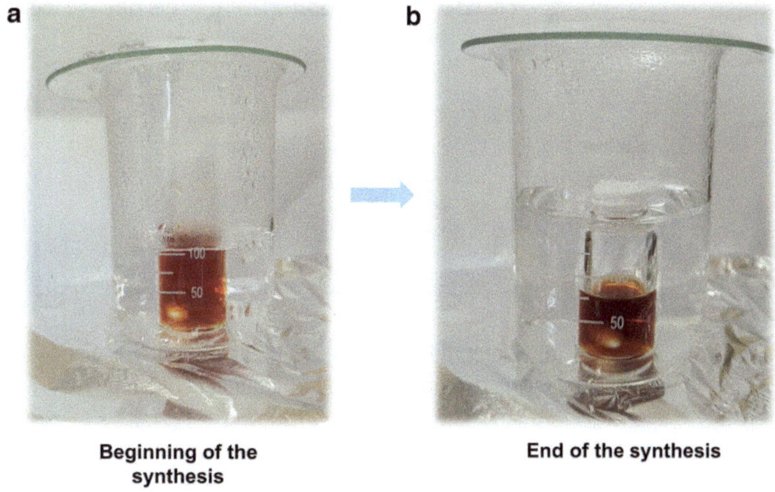

Fig. 2 Photographs of the CIS QDs core synthesis: (**a**) Beginning and (**b**) End

a widely used tactic because the use of the core-shell structure makes it possible to obtain materials with better characteristics.

1. To form the shell, add 40 µL of the zinc chloride solution to 5 mL of QD suspension.
2. Place the suspension in a water bath at approximately 100°C for 30 min.

3.1.3 Purification

Purification is used to remove excess of unreacted precursors.

1. Add acetone to the QDs until the mixture becomes cloudy.
2. Centrifuge at 4650 *RCF* for 10 min, and collect the precipitate containing the QDs.
3. Redisperse the precipitate in ultrapure water.
4. For analyses that require QDs in powder form, after centrifugation, the supernatant should be discarded, and the precipitate should be dried in an oven at 60°C for 1 h.
5. Fig. 3 shows the color variation of the resulting synthesis; it is possible to verify the increase in fluorescence after ZnS addition.

3.2 Characterization of QDs Using UV-Vis Absorption and Emission Spectroscopy

The synthesized materials are characterized optically by UV-vis absorption and photoluminescence spectroscopy. The absorption and emission spectra are shown in Fig. 4.

1. Diluting the sample may be necessary for absorbance and emission measurements (*see* **Note 4**).

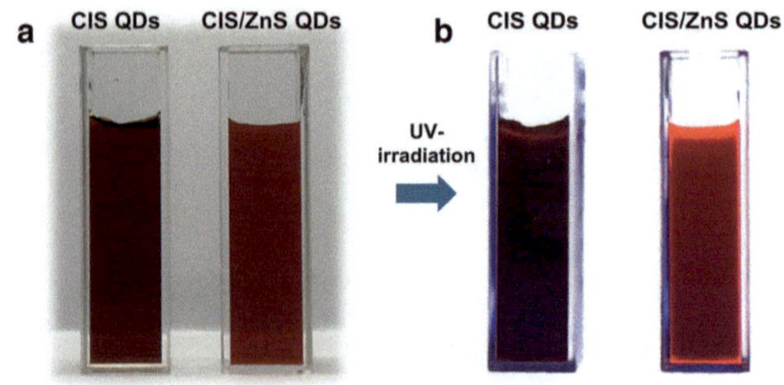

Fig. 3 (**a**) Photographs of the CIS and CIS/ZnS QDs under ambient light and (**b**) Fluorescence image under UV irradiation at 365 nm

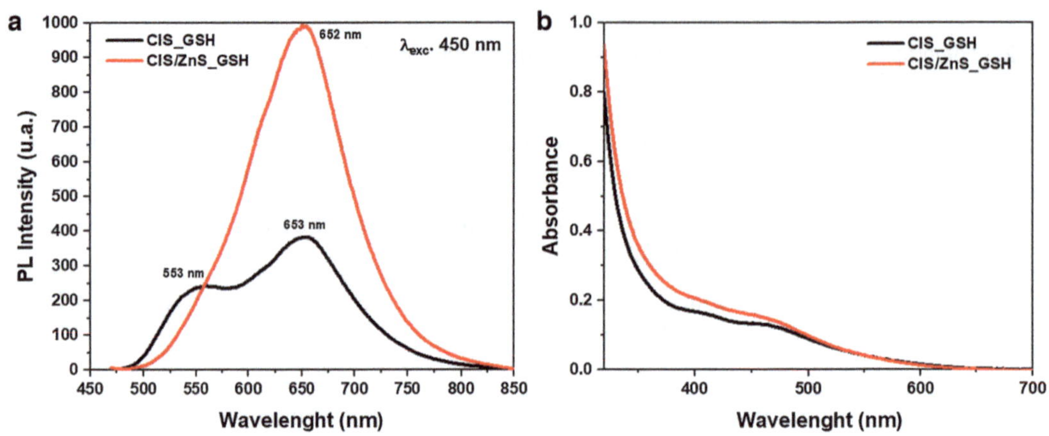

Fig. 4 **(a)** Photoluminescence spectra of the CIS and CIS/ZnS QDs **(b)** UV/Vis spectra of the CIS and CIS/ZnS QDs in aqueous medium

2. Measure the absorption of quartz cells with a 1 cm optical path in the 470–800 nm region.
3. The absorption and emission spectra show the formation of a photoluminescent material that absorbs and emits in the visible region of the electromagnetic spectrum. The material is good at absorbing light over a wide range of visible wavelengths.
4. Acquire the fluorescence spectra with excitation at 450 nm.
5. The materials present emission in the orange region, with a maximum of 652 nm. The increase in the relative emission intensity indicates the effectiveness of the formation of the shell due to the improvement in the relative emission properties of the synthesized nanocrystals, corroborating the change indicated in the emission spectra. The results corroborate the observations in Fig. 3.

6. The photoluminescence spectrum of the CIS QDs show a lower intensity band at 553 nm, which may be related to unreacted In_2S_3 or ultrasmall CIS QDs [14].

3.3 Characterization of QDs Using Time-Resolved Photoluminescence Spectra and XRD

3.3.1 Time-Resolved Photoluminescence

Time-resolved photoluminescence (TRPL) is defined as the average time a molecule spends in the excited state before returning to the ground state.

1. Stimulate the excitation using a nanoLED with a wavelength of 455 nm for CIS and CIS/ZnS QDs (*see* **Note 5**).
2. Fit the TRPL according to a bi-exponential function, shown in Eq. 1:

$$I(t) = A_1 exp(-t/\tau_1) + A_2 exp(-t/\tau_2) \quad (1)$$

   Where τ_1 and τ_2 are the decay times, and A_1 and A_2 are the fractional contributions of the two components.

3. Calculate the average decay time $\tau_{average}$ using Eq. 2:

$$\tau_{average} = \left(A_1\tau_1^2 + A_2\tau_2^2\right) / \left(A_1\tau_1 + A_2\tau_1\right) \quad (2)$$

4. The curves resulting from this adjustment for the CIS and CIS/ZnS QDs are presented in Fig. 5a.

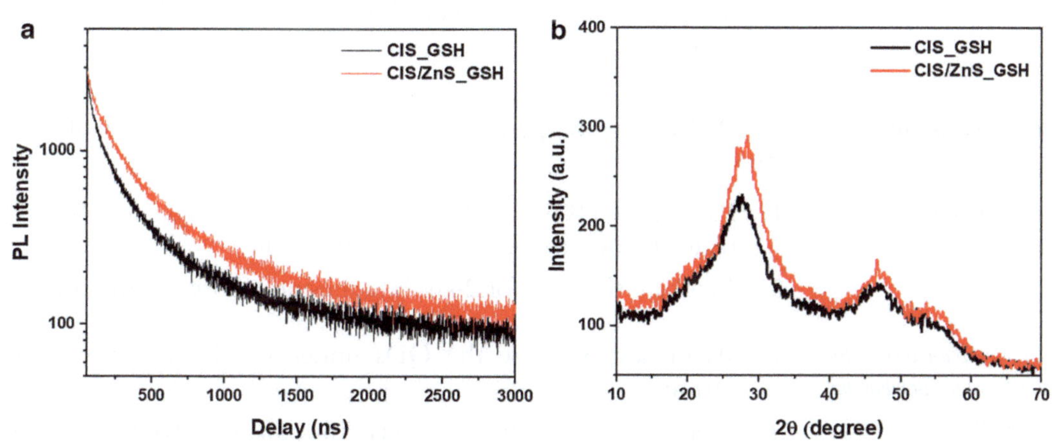

Fig. 5 (**a**) Fluorescence decay curves of the CIS and CIS/ZnS QDs (**b**) X-ray diffractograms of the CIS and CIS/ZnS QDs

5. The faster decay component (τ_1) represents the recombination of the electron/hole pair (radiative decay), and the longer component (τ_2) is associated with the decay inherent to defects existing on the surface of the material (trap states) (nonradiative decay) [15].
6. The decay times τ_1 and τ_2 were 52.66 and 358.58 ns for CIS QDs and 96.15 and 474.76 ns for CIS/ZnS QDs. The amplitudes A_1 and A_2 were 28.29 and 71.71 for the CIS QDs and 29.68 and 70.32 for the CIS/ZnS QDs.
7. The $\tau_{average}$ was 341.83 and 444.95 ns for CIS and CIS/ZnS, respectively.
8. The long lifetimes of hundreds of nanoseconds for the QDs synthesized in this work provide an attractive alternative for cell labeling through time-gated detection (TGD). In this technique, when an excitation pulse is performed, the short-lived emitting components of the biological environment will soon no longer be emitting, while the luminescence of the long-lived components (CIS/ZnS QDs) will still be present [11].

3.3.2 X-Ray Diffraction (XRD)

X-ray diffraction (XRD) analyses were performed in order to better understand the structure and composition of the crystalline phases present in the QDs (*see* **Note 6**). The diffractograms are presented in Fig. 5b.

1. The peaks of the diffraction pattern of the CIS ($2\theta = 27.4$, 46.8 and 54.29°) and CIS/ZnS ($2\theta = 28.2$, 47.2 and 54.5°) QDs are related to the (112), (204), and (312) planes, corresponding to the tetragonal phase of chalcopyrite (JCPDS card no. 01-085-1575), as shown in Fig. 5b.
2. No other impurity phases were detected in the XRD spectrum, indicating that highly crystalline CIS and CIS/ZnS QDs were prepared.

3.4 MTT Assay

The MTT assay was performed to evaluate the cytotoxicity of CIS/ZnS QDs.

3.4.1 Cell Culture Preparation

1. Plate the CHO-K1 cells (after the third passage) in a 96-well plate at a density of 1×10^4 cells per well.
2. Incubate the cells for 24 h for adhesion before treatment.

3.4.2 Preparation of CIS/ZnS QDs Stock Solution for MTT Assays

1. Weigh and sterilize the QDs under UV light exposure for 20 min.
2. Prepare the test concentrations (150, 80, and 20 μg/mL) from the stock concentration in DMEM medium (*see* **Note 7**).

3.4.3 Experimental Treatment

1. Treat the cells with CIS/ZnS QDs at concentrations of 150, 80, and 20 μg/mL.
2. To validate the experiment, six wells are required for each concentration studied.
3. Perform a sextuplicate of the control (cells without the presence of QDs) and blank (absence of cells, containing only DMEM).

3.4.4 MTT Assay Protocol

After 24 h of treatment with CIS/ZnS QDs, the MTT assay is conducted according to the protocol described.

1. Remove 100 μL of the medium from each well.
2. Add 10 μL of the MTT solution to each well.
3. Incubate for 4 h at 37°C to allow formation of formazan crystals.
4. Add 100 μL of SDS to promote the lysis of formazan crystals.
5. Shake for 60 min for complete solubilization.
6. Measure the absorbance using an ELISA reader with a 546 nm filter.

3.4.5 MTT Assay Results

1. As illustrated in Fig. 6, the CIS/ZnS QDs does not cause serious cytotoxicity and exhibits high cell viability even at concentrations up to 150 μg/mL.
2. The results show that CIS/ZnS-GSH QDs exhibit extremely low toxicity and excellent biocompatibility.

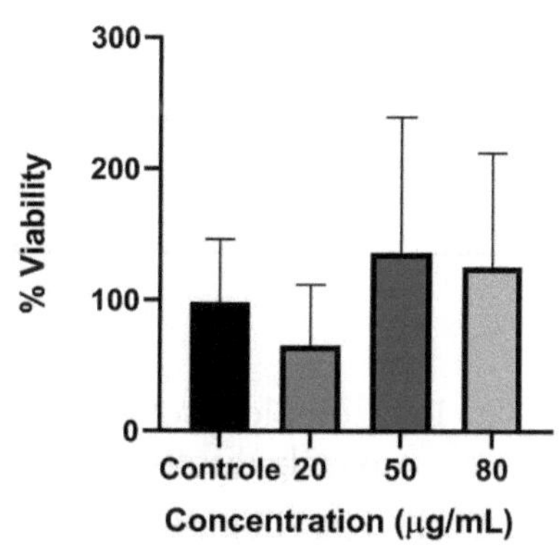

Fig. 6 Cell viability of CHO cells incubated with various concentrations of CIS/ZnS QDs. The MTT assay was carried out for each QDs concentration in sextuples. Viability was tested by analysis of variance, and CIS/ZnS QDs did not exhibit cytotoxicity ($p > 0.05$)

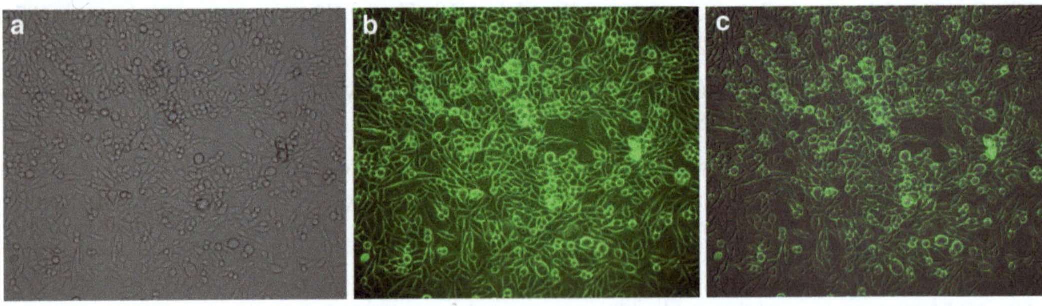

Fig. 7 Fluorescence images of CHO cells (**a**) bright field, (**b**) fluorescence field, and (**c**) merged images

3.5 Live Cell In Vitro Imaging

1. Incubate the cells in 12-well plates (5×10^5 per well) for 3 days until adherence.
2. Aspire the medium and thoroughly wash the plates with PBS.
3. Incubate the plates with CIS/ZnS QDs for 15 min at a concentration of 150 μg/mL.
4. Place the cells under a fluorescence microscope equipped with the solid-state LED lamp illumination system and choose the blue LED (excitation 470 nm) to check the emission of the QDs in the cells.
5. Image processing is done in ImageJ 1.53t (Fig. 7). Add the fluorescence image to the software and click Adjust > Color Balance > Select Green and adjust the minimum, maximum, and brightness to improve the image contrast. To merge the bright field images with the fluorescence image, select Image > Color > Merge Channels.
6. The cell morphologies are clearly observed.
7. The high fluorescence intensity in the plasma membrane was due to the abundance of QDs in this region, suggesting that QDs are capable of penetrating cell membranes to enter cells.

4 Notes

1. Complete solubilization of GSH requires stirring until completely dissolved.
2. The solution turns whitish at this stage of preparation.
3. Place a watch glass in the beaker to prevent heat loss.
4. It is necessary to dilute until an absorbance of less than one is found to comply with Beer-Lambert law.
5. The selection of a nanoLED depends on the excitation wavelength.

6. For XRD analysis, the purified sample was oven-dried at 60°C for 1 h before being placed in a specific sample holder for XRD. The diffraction pattern was obtained using a diffractometer with radiation, Kα of Cu, $\lambda = 1.5418$ Å, graphite monochromator and sweep of 0.02 degrees, between $2\theta = 10$–80°).
7. For complete dispersion of the QDs, it may be necessary to use ultrasound to disperse the entire sample.

Acknowledgments

The authors would like to thank the financial support from (FAPEMIG) Fundação de Amparo à Pesquisa do Estado de Minas Gerais), CNPq (Conselho Nacional de Desenvolvimento Científico e Tecnológico), CAPES (Coordenação de Aperfeiçoamento de Pessoal de Nível Superior), and FINEP (Financiadora de Estudos e Projetos).

References

1. Ekimov AI, Onushchenko AA (1981) Quantum size effect in three-dimensional microscopic semiconductor crystals. JETP Lett 118:S15–S17
2. Wegner KD, Hildebrandt N (2024) Near infrared quantum dots for biosensing and bioimaging. Trends Anal Chem 180:117922. https://doi.org/10.1016/j.trac.2024.117922
3. Sobhanan J, Rival JV, Anas A et al (2023) Luminescent quantum dots: synthesis, optical properties, bioimaging and toxicity. Adv Drug Deliv Rev 197:114830. https://doi.org/10.1016/j.addr.2023.114830
4. Wu J, Li J, Li L et al (2024) Water-soluble near-infrared CuInS2/ZnS fluorescent quantum dots for Cu^{2+} detection and cellular imaging. Microchem J 199:109962. https://doi.org/10.1016/j.microc.2024.109962
5. Sun H, Zhou P, Su B (2023) Electrochemiluminescence of semiconductor quantum dots and its biosensing applications: a comprehensive review. Biosensors (Basel) 13:708. https://doi.org/10.3390/bios13070708
6. Medina-Berríos N, Pantoja-Romero W, Lavín Flores A et al (2023) Synthesis and characterization of carbon-based quantum dots and doped derivatives for improved andrographolide's hydrophilicity in drug delivery platforms. ACS Omega 9:12575–12584. https://doi.org/10.1021/acsomega.3c06252
7. Zorab MM, Mohammadjani N, Ashengroph M, Alavi M (2023) Biosynthesis of quantum dots and their therapeutic applications in the diagnosis and treatment of cancer and SARS CoV-2. Adv Pharm Bull 13:411–422. https://doi.org/10.34172/apb.2023.065
8. Neethu KM, Karmakar S, Sahoo B et al (2025) Use of quantum dots as nanotheranostic agents: emerging applications in rare genetic diseases. Small 2407353:1–8. https://doi.org/10.1002/smll.202407353
9. Zhang Y, Huang J, Ding Q et al (2025) One step in situ synthesis of Cu/Mn: ZnInSe quantum dots with enhanced luminescence and environmental stability. Spectrochim Acta A Mol Biomol Spectrosc 333:125842. https://doi.org/10.1016/j.saa.2025.125842
10. Zhang S, Tang X, Zang L, Zhao L (2024) Carbon quantum dots(CQDs)-sensitized CdS/$CuInS_2$ heterojunction as a photoelectrochemical biosensing platform for highly sensitive detection of prostate-specific antigen. Talanta 272:125811. https://doi.org/10.1016/j.talanta.2024.125811
11. Morselli G, Villa M, Fermi A et al (2021) Luminescent copper indium sulfide (CIS) quantum dots for bioimaging applications. Nanoscale Horiz 6:676–695. https://doi.org/10.1039/d1nh00260k
12. Lv Y, Zhang L, Wu R, Li LS (2024) Recent progress on eco-friendly quantum dots for bioimaging and diagnostics. Nano Res 17:10309–10331. https://doi.org/10.1007/s12274-024-6926-5
13. Chen T, Chen Y, Li Y et al (2023) A review on multiple I-III-VI quantum dots: preparation and enhanced luminescence properties. Materials

16:5039.https://doi.org/10.3390/ma16145039
14. Yuan Z, Yang L, Han D et al (2021) Synthesis and optical properties of In2S3-hosted colloidal Zn-Cu-In-S nanoplatelets. ACS Omega 6:18939–18947. https://doi.org/10.1021/acsomega.1c02180
15. Hashemkhani M, Loizidou M, Macrobert AJ, Yagci Acar H (2022) One-step aqueous synthesis of anionic and cationic AgInS2 quantum dots and their utility in improving the efficacy of ALA-based photodynamic therapy. Inorg Chem 61:2846–2863. https://doi.org/10.1021/acs.inorgchem.1c03298

Check for updates

Chapter 5

Preparation of Bimodal Nanoprobes for Fluorescent and Magnetic Resonance Imaging by Conjugating Gadolinium Complexes to Quantum Dots

Gabriela M. Albuquerque, Rebeca M. Melo, Samantha D. Coiado, Carlos F. G. C. Geraldes, Giovannia A. L. Pereira and Goreti Pereira

Abstract

The development of strategies to optimize the conjugation of Gd^{3+} complexed to quantum dots (QDs) has been increasingly explored. The aim is to combine two imaging techniques, improving diagnostics and overcoming the limitations of individual methods. This chapter describes the preparation of bimodal systems by conjugating GdDOTA-thiolated complexes with fluorescent QDs through dative bonds, in an aqueous medium. Due to the toxicity of the Gd^{3+} ion, structural modifications of the DOTA-based ligands were evaluated to ensure that these changes did not compromise the complexation efficiency or the stability of the resulting complexes. The optical and relaxometric properties of the conjugated complexes to the QDs were also assessed.

Keywords Paramagnetic complex, DOTA, Dative bond, Xylenol orange test, Transmetallation, Semiconductor nanocrystals

1 Introduction

In recent years, the conjugation of Gd^{3+} complexes to quantum dots (QDs) has been explored as a strategy to combine the paramagnetic advantages of Gd^{3+} complexes with the optical properties of QDs [1, 2]. In addition, the attachment of a larger entity to the complex can influence its relaxometric features, enhancing its efficiency as a contrast agent (CA) for magnetic resonance imaging (MRI) [3]. Gd^{3+} complexes are among the most prominent T_1-based CAs used clinically, owing to the ion's excellent paramagnetic properties, having nine sites available for coordinated binding. Furthermore, Gd^{3+} exhibits a relatively long electronic relaxation time, which contributes to its high efficacy in MRI applications [4, 5].

In turn, QDs are semiconductor nanocrystals that stand out for their unique optical properties, and highly reactive surface suitable for

Goreti Pereira et al. (eds.), *Quantum Dots*, Methods in Molecular Biology, vol. 3025,
https://doi.org/10.1007/978-1-0716-5210-7_5,

conjugation. There are several strategies for attaching molecules to QDs, including noncovalent interactions, dative bonds, and covalent binding. Conjugation by dative bonds involves coordination linkage between the QD metal and a thiol group present in the molecule of interest, forming a bond with less energy than a covalent bond [1, 6, 7].

The QD colloidal suspensions can be prepared in either organic or aqueous media, and are typically formed through the reaction of a metal precursor, a chalcogenide source, stabilizer molecule, and reducing agent, under controlled experimental conditions, such as pH and temperature [8]. Given the potential biomedical applications of these nanocrystals, their preparation in aqueous media offers a distinct advantage, as it yields hydrophilic QDs without requiring further surface modification steps. Furthermore, this approach reduces the environmental impact of the overall synthesis process [9].

This chapter describes the preparation of three Gd^{3+} complexes containing thiol groups and their association with CdTe QDs, to achieve bimodal nanoprobes using a practical and versatile conjugation strategy. It also outlines the methods used to assess the complex formation efficiency and stability, and the bimodal systems optical and relaxometric properties. The Gd^{3+} complexes association with the QDs through the metal-thiol bond described here is based on the methodology presented by Albuquerque et al. [10].

2 Materials

Prepare all solutions using ultrapure water (18.2 MΩ.cm^{-1} at 25°C) and store all prepared solutions in a refrigerated environment.

2.1 GdDOTA Complexes

1. NaOH solution, 0.02 mol/L: dissolve 200 mg (0.005 mol) of NaOH to 250 mL of ultrapure water.
2. 1,4,7,10-Tetraazacyclododecane-1,4,7,10-tetraacetic acid (DOTA) solution, 0.014 mol/L: dissolve 330 mg (0.5 mmol) of DOTA in 35 mL of ultrapure water.
3. *N*-ethyl-*N*'-(3-dimethylaminopropyl) carbodiimide (EDC) solution, 0.1 mol/L: dissolve 96 mg (0.5 mmol) of EDC in 5 mL of ultrapure water.
4. *N*-hydroxysuccinimide (NHS) solution, 0.1 mol/L: dissolve 58 mg (0.5 mmol) of NHS in 5 mL of ultrapure water.
5. Gd^{3+} solution, 0.094 mol/L: dissolve 123.90 mg (0.47 mmol) of $GdCl_3$ in 5 mL of ultrapure water.

2.2 Xylenol Orange

1. Acetic acid buffer solution, 0.05 mmol/L: dilute 720 μL of acetic acid in 200 mL of ultrapure water and adjust the pH to 5.8 (*see* **Note 1**). Adjust the final volume to 250 mL with ultrapure water.

2. Xylenol orange solution, 0.016 mM: dissolve 3 mg of xylenol orange in 250 mL of the acetic acid buffer solution. Store at −6°C (*see* **Note 2**).
3. Gd^{3+} solution, 0.001 mol/L: dissolve 5.27 mg of $GdCl_3$ (0.02 mmol) in 20 mL of ultrapure water.

2.3 Transmetallation

1. Phosphate buffer solution, 0.067 mol/L: add 5.83 mg of Na_2PO_4 (0.041 mmol) and 3.54 mg de KH_2PO_4 (0.026 mmol) in 10 mL of ultrapure water. Adjust the pH to 7 (*see* **Note 1**).
2. Zinc chloride solution, 250 mmol/L: add 34.1 mg (0.3 mmol) of $ZnCl_2$ in 1 mL of ultrapure water.
3. Dilute 25 µL of each complex solution (prepared in Sect. 3.1) in 975 µL of phosphate buffer solution to obtain solutions with a final concentration of 2.5 mmol/L.

2.4 Synthesis of CdTe QDs

1. NaOH solution (2.0 mol/L) in ultrapure water
2. Sodium tellurite solution (Na_2TeO_3), 0.02 mol/L: dissolve 132.9 mg (0.60 mmol) of Na_2TeO_3 in 30 mL of ultrapure water.
3. Sodium borohydride ($NaBH_4$) solution, 2.5 mol/L: dissolve 94.6 mg (2.5 mmol) of $NaBH_4$ in 1 mL of ultrapure water.
4. Sodium borohydride ($NaBH_4$) solution, 0.6 mol/L: dissolve 113.5 mg (3 mmol) of $NaBH_4$ in 5 mL of ultrapure water.

2.5 Optical Measurements

1. UV-Vis absorption spectrophotometry: dilute the samples at 1:10 ratio (sample to final volume).
2. Emission spectrometer: dilute the samples at 1:10 ratio (sample to final volume).

2.6 Relaxivity Measurements

1. Relaxometer equipment operating at 0.49 T (corresponding to 20 MHz proton Larmor frequency, at 25°C and 37°C) and at 1.4 T (corresponding to 60 MHz proton Larmor frequency, at 37°C), equipped with a temperature control system.

3 Methods

3.1 GdDOTA Complexes

1. Use 35 mL of the DOTA solution (0.014 mol/L) and adjust the pH to 5.6–5.8 (*see* **Note 3**).
2. Add 5 mL of EDC solution (0.1 mol/L) and 5 mL of NHS solution (0.1 mol/L).
3. Leave the system under stirring for 1 h at 60°C.
4. With the solution at room temperature (~25°C), adjust the pH to 7.

5. Add 5 mL of Gd^{3+} solution (0.094 mol/L) to the solution containing DOTA (*see* **Note 4**).
6. Leave the final solution under stirring for 1 h at 60°C.
7. At room temperature, readjust the pH to 7, and heat again at 60°C for 30 min. Repeat this step until the pH remains close to 7 (*see* **Note 5**).
8. Store the GdDOTA-NHS solution at 4°C until use.
9. Linker addition: add 0.05 mmol of the chosen linker—cysteamine (CYA), L-cysteine (Cys), or glutathione (GSH)—to 5 mL of GdDOTA-NHS solution (Fig. 1).
10. Leave under stirring for 4 h at room temperature (~25°C). Adjust the final pH to 7.

3.2 Xylenol Orang Test

The xylenol orange tests are performed based on that proposed by Barge [11].

1. To achieve the calibration curve, prepare a series of xylenol and Gd^{3+} solutions. To 2 mL of 0.016 mmol/L xylenol orange solution, add predefined volumes (0, 1, 2, 4, 6, 8, 10 μL) of a 0.001 mol/L Gd^{3+} solution.
2. Acquire the UV-Vis absorption spectra for each xylenol—Gd^{3+} solution, in the range of 300–700 nm.
3. Plot a graph of the added Gd^{3+} concentration (in μmol/L) as a function of absorbance ratio at 573 and 433 nm. Perform a linear regression to obtain the calibration curve equation (Eq. 1, Fig. 2).

$$\frac{A_{573}}{A_{473}} = m\,[Gd^{3+}] + b \qquad (1)$$

4. To analyze the GdDOTA complexes prepared in Sect. 3.1, add varying volumes of the complex solution (0, 1, 2, 3, 4,

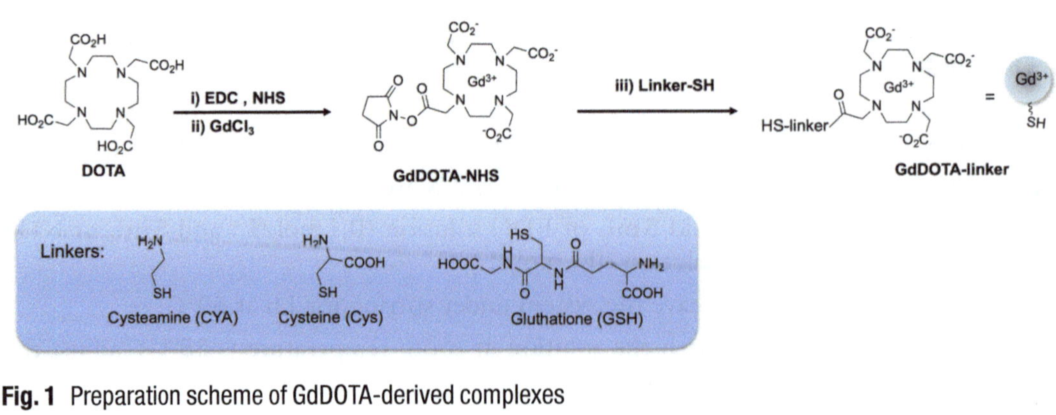

Fig. 1 Preparation scheme of GdDOTA-derived complexes

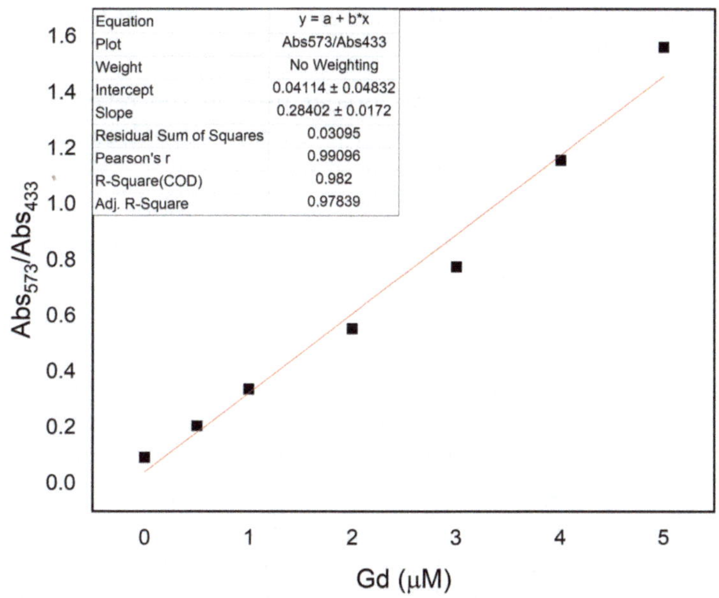

Fig. 2 Graph obtained by linear regression based on the points obtained in the calibration curve of the xylenol orange test, associating the concentration of Gd^{3+} with the ratio between absorbances at 573 and 433 nm

and 5 μL) to 2 mL of xylenol orange solution (0.016 mM). Acquire the UV-Vis absorption spectra of each mixture over the range of 300–700 nm.

5. Using the absorption values at the reference wavelengths (573 and 433 nm) and the calibration curve, estimate the concentration of free Gd^{3+} in solution. Based on this, determine the complexation yield for each GdDOTA complex.

3.3 GdDOTA Complexes Relaxometric Characterization

1. Add 1 mL of the GdDOTA complex solution to the relaxometer sample holder.
2. Measure the longitudinal relaxation time (T_1) in the relaxometer at controlled temperature conditions (*see* **Note 6**).
3. In this case, the diamagnetic contribution ($T_{i,d}$) corresponds to the relaxation time measured for ultrapure water, while the measurements of the complexes represent the observed relaxation time ($T_{i,obs}$).
4. Calculate the relaxivity using Eq. (2), as proposed by Caravan (2009) [12], taking into account the previously determined Gd^{3+} concentration in each complex solution.

$$r_1 = \frac{\left(\frac{1}{T_1}\right)_p}{[Gd^{3+}]} = \frac{\left(\frac{1}{T_1}\right)_d - \left(\frac{1}{T_1}\right)_{obs}}{[Gd^{3+}]} \tag{2}$$

3.4 Transmetallation Assay

The transmetallation tests are carried out based on that proposed by Laurent [13].

1. In the sample holder used in the relaxometer, add 1 mL of the solution of the complex dissolved in phosphate buffer, described in Sect. 2.3.
2. Quickly, add 10 μL of zinc chloride solution (250 mmol/L) to the sample holder and measure immediately the T_1 on the relaxometer (measurement at 0 min), using the same parameters as in Sect. 3.3.
3. Measure the T_1 relaxation time measurements every 20 min during the first hour, using the same parameters as in Sect. 3.3. After the first hour, perform measurements every 24 h, until a total duration of 7200 min (5 days) is completed.
4. Analyze the results by evaluating the evolution of the longitudinal paramagnetic relaxation rate ($R_{1p} = 1/T_{1p}$) of the ^{1}H nuclei of the water molecules over time (Fig. 3). The ratio $R_1(t)/R_1(0)$ should remain greater than 0.99 over 5 days, indicating that the longitudinal relaxation time does not increase over time, and consequently, there is no decomplexation of Gd^{3+}.
5. Determine the thermodynamic $[R_1(t)/R_1(0)]$ and kinetic (time for $R_1(t)/R_1(0) = 0.8$) indices.

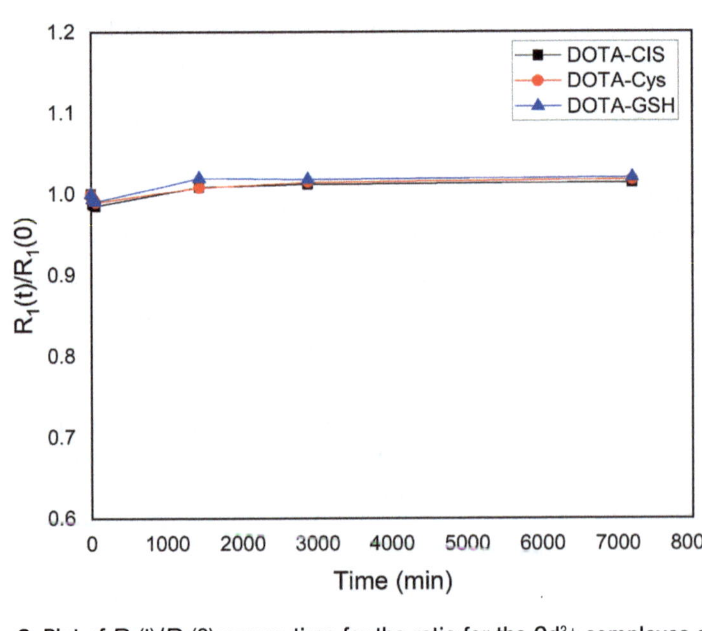

Fig. 3 Plot of $R_1(t)/R_1(0)$ versus time for the ratio for the Gd^{3+} complexes over 5 days (7200 min)

3.5 Synthesis of CdTe QDs

This is based on the one-pot method presented by Albuquerque et al. [10], to prepare CdTe QDs by using two different stabilizing agents, namely: CYA and GSH.

3.5.1 Synthesis of CdTe-CYA

1. In a two-necked flask, dissolve 4.7 mmol of $CdCl_2$ and 8.4 mmol of CYA in 125 mL of ultrapure water.
2. Adjust the pH to 5.8–6.0, using a NaOH solution (2.0 M).
3. Heat the system to 90°C, under stirring and N_2 gas flux for 30 min.
4. Add 25 mL of the Na_2TeO_3 solution (0.02 mol/L).
5. Add 1 mL of the $NaBH_4$ solution (2.5 mol/L) (*see* **Note 7**).
6. Maintain the reaction stirring at 90°C and under a N_2 flux for 30 min.
7. Turn off the gas, close carefully the system, and leave it under heating and stirring for 5 h.
8. Let the synthesis temperature decrease until room temperature (~25°C), and adjust the pH to 5.6–5.8 by adding CYA (*see* **Note 8**).
9. Finally, store the synthesis under refrigeration (~5°C).

3.5.2 Synthesis of CdTe-GSH

1. In a two-necked flask, dissolve 0.20 mmol of $CdCl_2$ and 0.24 mmol of GSH in 80 mL of ultrapure water.
2. Adjust the pH to 7.0–8.0, using a NaOH solution (2.0 mol/L).
3. Heat the system to 90°C, under stirring and N_2 gas flux for 30 min.
4. Add 5 mL of the Na_2TeO_3 solution (0.02 mol/L).
5. Add 5 mL of the $NaBH_4$ solution (0.6 mol/L) (*see* **Note 7**).
6. Maintain the reaction stirring at 90°C and under a N_2 flux for 30 min.
7. Turn off the gas, close carefully the system, and leave it under heating and stirring for 5 h.
8. Finally, store the synthesis under refrigeration (~5°C).

3.6 Conjugation of the GdDOTA Complex With QDs

1. Centrifuge the QDs at ~805 × *g* for 3 min using an ultracentrifuge tube with a 10 kDa MWCO membrane (*see* **Notes 9 and 10**).
2. Dilute 0.5 mL of the centrifuged QD suspension in 1.5 mL of ultrapure water.
3. Add 15.6 and 26.1 mL of the complex solution (0.01 mol/L) to CdTe-CYA and CdTe-GSH, respectively (*see* **Note 11**) (Fig. 4).
4. Stir for 4 h at room temperature and store under refrigeration (~5°C).

5. Evaluate the effect of conjugation on the optical properties of QDs by UV-Vis absorption and emission spectroscopy (Fig. 5).

3.7 QDs-GdDOTA Systems Optical Characterization

1. Absorption measurements: perform analyses in the wavelength range of 300–700 nm.
2. Emission measurements: excite the sample at the wavelength of the first absorption maximum (*see* **Note 12**).

3.8 QDs-GdDOTA Systems Relaxometric Characterization

1. Centrifuge the QDs-GdDOTA systems at ~805 × *g* for 3 min, using ultracentrifuge tubes with 10 kDa MWCO membrane (*see* **Note 13**).
2. Perform measurements of the longitudinal (T_1) and transverse (T_2) water proton relaxation times (*see* **Note 6**).
3. In this case, the diamagnetic contribution ($T_{i,d}$) is the water proton relaxation time measured for the aqueous suspension containing only the QD, while the measurements of the systems containing the QDs-GdDOTA correspond to $T_{i,obs}$.
4. Determine the Gd^{3+} concentration in each system analyzed (*see* **Note 14**).
5. Calculate the longitudinal (r_1) and transverse (r_2) relaxivities using Eq. (2).

4 Notes

1. The pH adjustment is done using a 0.02 mol/L NaOH solution.

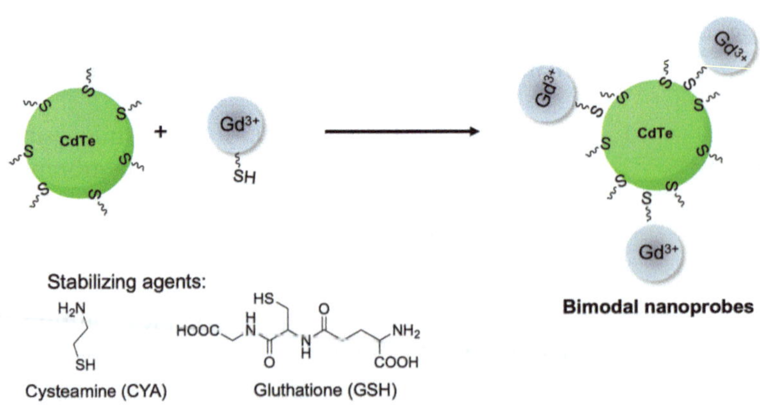

Fig. 4 Scheme of the noncovalent conjugation of complexes directly on the QD surface, adapted from Albuquerque (2024) (reproduced with permission of [10])

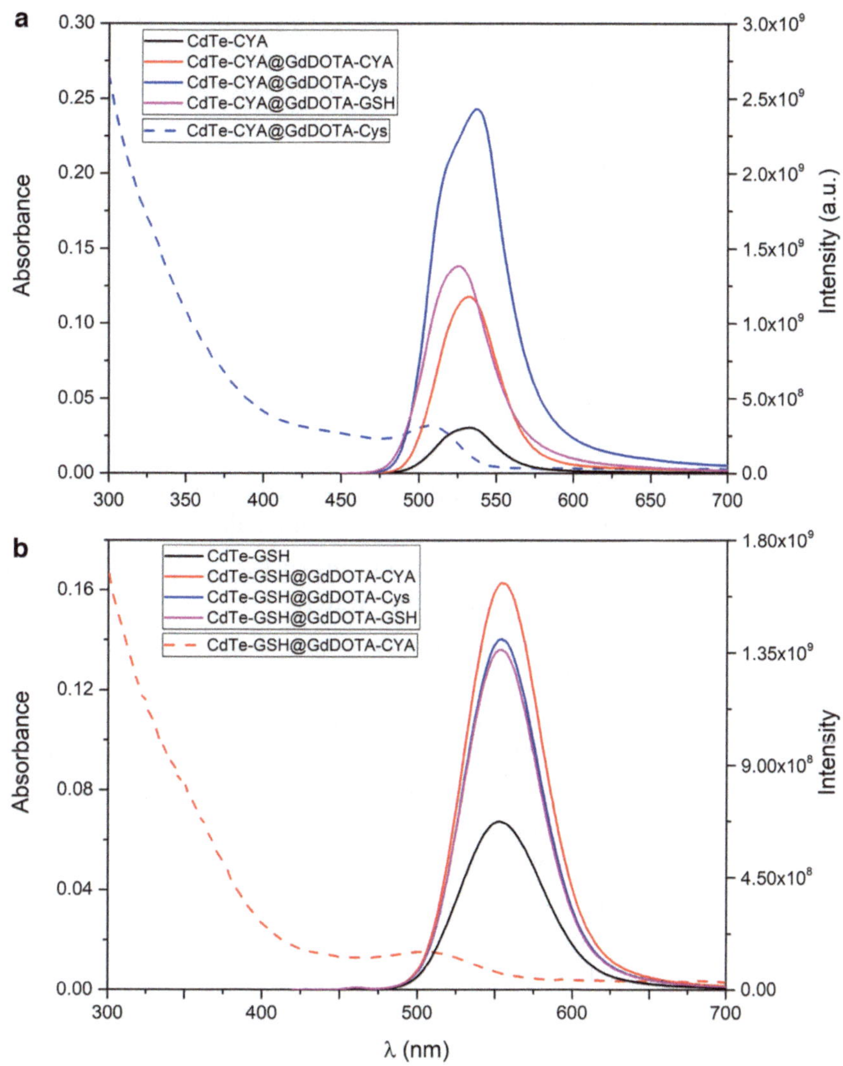

Fig. 5 Optical characterization of bimodal systems prepared by metal-thiol conjugation of CdTe QDs functionalized with **(a)** CYA and **(b)** GSH, with the GdDOTA-derived complexes. Absorption spectrum (dashed line) of one of the systems to exemplify the profile of the other systems. Emission spectra (solid line) of the systems prepared with the three different complexes

2. After being stored at low temperature and once brought back to room temperature, the xylenol orange solution must not be refrigerated again, and should be used within 45 min after thawing [11].
3. The pH is adjusted to ensure optimal pH conditions for the reaction with EDC [6].
4. The Gd^{3+} amount used for complexation should be 95% relatively to the DOTA ligand, in order to minimize the presence of uncomplexed Gd^{3+} ions in solution.

5. After every 30 min of stirring at 60°C, the solution is allowed to return to room temperature to readjust the pH to 7. This procedure is repeated until, after stirring under heating, the pH remains stable at 7.
6. We used a relaxometer operating at 0.49 and 1.4 T. A nuclear magnetic resonance spectrometer operating at any magnetic field strength may also be employed. However, it is essential to report the relaxivity values along with the corresponding measurement frequency, as relaxivity is dependent on the magnetic field strength.
7. The $NaBH_4$ solution should be freshly prepared before the synthesis.
8. This will prevent the synthesis from precipitating.
9. To remove impurities from the synthesis, such as possible free Cd^{2+} ions and stabilizer molecules.
10. After centrifugation, return the sample to the initial volume by adding ultrapure water.
11. Each system was prepared using a QD to complex ratio of 1:30.
12. For emission measurements, the systems presented in the work described in this chapter were excited at 488 nm.
13. Centrifugation plays a key role in removing free Gd^{3+} ions and unbound complexes in solution, which affect the accuracy of relaxation time measurements.
14. Determine the Gd^{3+} concentrations in each sample using an inductively coupled plasma–optical emission spectroscopy (ICP-OES).

Acknowledgments

This work was supported by Fundação para a Ciência e a Tecnologia (FCT) I.P. under the project/grant UID/50017/2025 (doi.org/10.54499/UID/50017/2025) and LA/P/0094/2020 (doi.org/10.54499/LA/P/0094/2020), CNPq (409319/2021-0), and FACEPE (APQ-1351-1.06/22 and BFP-0119-3.07/24).

References

1. Albuquerque GM, Souza-Sobrinha I, Coiado SD, Santos BS, Fontes A, Pereira GAL et al (2021) Quantum dots and Gd^{3+} chelates: advances and challenges towards bimodal nanoprobes for magnetic resonance and optical imaging. Top Curr Chem 379:12. Springer Science and Business Media Deutschland GmbH. https://doi.org/10.1007/s41061-021-00325-x
2. Li Y, Zhang P, Tang W, Mchugh KJ, Kershaw SV, Jiao M et al (2022) Bright, magnetic NIR-II

quantum dot probe for sensitive dual-modality imaging and intensive combination therapy of cancer. ACS Nano 16:8076–8094. https://doi.org/10.1021/acsnano.2c01153
3. MacMahon E, Brougham DF (2023) pH dependence of MRI contrast in magnetic nanoparticle suspensions demonstrates inner-sphere relaxivity contributions and reveals the mechanism of dissolution. Langmuir 39(6):2171–2181. https://doi.org/10.1021/acs.langmuir.2c02621
4. Oliveira AC, Filipe HAL, Ramalho JPP, Salvador A, Geraldes CFGC, Moreno MJ et al (2022) Modeling Gd^{3+} complexes for molecular dynamics simulations: toward a rational optimization of MRI contrast agents. Inorg Chem 61(30):11837–11858. https://doi.org/10.1021/acs.inorgchem.2c01597
5. Oliveira AC, Filipe HAL, Geraldes CFGC, Voth GA, Moreno MJ, Loura LMS (2024) Interaction of MRI contrast agent $[Gd(DOTA)]^-$ with lipid membranes: a molecular dynamics study. Inorg Chem 63:10897–10914. https://doi.org/10.1021/acs.inorgchem.4c00972
6. Pereira G, Monteiro CAP, Albuquerque GM, Pereira MIA, Cabrera MP, Cabral Filho PE et al (2019) (Bio)conjugation strategies applied to fluorescent semiconductor quantum dots. J Braz Chem Soc. Sociedade Brasileira de Quimica 30:2536–2560. https://doi.org/10.21577/0103-5053.20190163
7. Fatima I, Rahdar A, Sargazi S, Barani M, Hassanisaadi M, Thakur VK (2021) Quantum dots: synthesis, antibody conjugation, and HER2-receptor targeting for breast cancer therapy. J Funct Biomater 12(4):75. https://doi.org/10.3390/jfb12040075
8. Mrad R, Poggi M, Ben Chaâbane R, Negrerie M (2020) Role of surface defects in colloidal cadmium selenide (CdSe) nanocrystals in the specificity of fluorescence quenching by metal cations. J Colloid Interface Sci 571:368–377. https://doi.org/10.1016/j.jcis.2020.03.058
9. Zhao JY, Wang ZG, Hu H, Zhang ZL, Tang B, Luo MY et al (2022) How different are the surfaces of semiconductor Ag_2Se quantum dots with various sizes? Sci Bull (Beijing) 67(6):619–625. https://doi.org/10.1016/j.scib.2021.12.004
10. Albuquerque GM, Melo RM, Coiado SD, Pereira GAL, Pereira G (2024) Nanoprobes based on quantum dots and Gd(III) complexes for dual optical and magnetic resonance imaging. Mater Lett [Internet] 371:136921. https://doi.org/10.1016/j.matlet.2024.136921
11. Barge A, Cravotto G, Gianolio E, Fedeli F (2006) How to determine free Gd and free ligand in solution of Gd chelates. A technical note. Contrast Media Mol Imaging 1(5):184–188 . https://doi.org/10.1002/cmmi.110
12. Caravan P, Farrar CT, Frullano L, Uppal R (2009) Influence of molecular parameters and increasing magnetic field strength on relaxivity of gadolinium- and manganese-based T1 contrast agents. Contrast Media Mol Imaging 4(2):89–100. https://doi.org/10.1002/cmmi.267
13. Laurent S, Van der Elst L, Henoumont C, Muller RN (2010) How to measure the transmetallation of a gadolinium complex. Contrast Media Mol Imaging 5(6):305–308. https://doi.org/10.1002/cmmi.388

Part II

Computational Tools

Chapter 6

Deep Learning Tools for Single Quantum Dot Tracking

Oleg Kovtun

Abstract

Deep learning, a powerful and widely effective subset of machine learning, offers a promising foundation for developing an automatic, parameter-free analysis pipelines that replace conventional, multistep workflows for single quantum dot tracking. This chapter provides detailed protocols for installing and implementing open-source, state-of-the-art deep learning tools to detect individual quantum dots in time-lapse image series and reconstruct their trajectories. Additionally, it includes practical guidelines for troubleshooting common errors encountered when deploying these tools for single quantum dot tracking.

Keywords Quantum dots, Single particle tracking, Machine learning, Deep learning, Convolutional neural network

1 Introduction

Single particle tracking (SPT) is a powerful biophysical technique first introduced using the term "digital video microscopy" 30 years ago. The pioneering work of Crocker and Greer involved time-lapse imaging of polystyrene sulfate spheres with a diameter of 1 μm on an inverted conventional light microscope to study time-resolved sphere interaction, complex fluid dynamics, and tissue mechanical properties at the single particle trajectory level in real time [1]. Steady improvements in instrumentation, probe biophysics, and computational power have effectively lowered SPT implementation barriers for nonexperts to record the dynamic motion of single molecules in living cells with high spatial and temporal precision [2, 3]. However, current limitations of the conventional, multistep SPT analysis pipeline stymie broader adoption of the SPT tools.

Traditionally, extracting quantitative information from the acquired movies of moving particles has relied on algorithmic approaches encompassing three distinct phases: (a) localization of single emitters with subpixel accuracy, (b) linkage of emitter positions in consecutive frames into trajectories, and (c) characterization of motion dynamics to gain meaningful insights into biological

Goreti Pereira et al. (eds.), *Quantum Dots*, Methods in Molecular Biology, vol. 3025,
https://doi.org/10.1007/978-1-0716-5210-7_6,

processes [3–5]. Inherent heterogeneity of imaging conditions and acquired data, including low signal-to-noise ratio (SNR), inhomogeneous and unsteady illumination, heterogeneous background, high object density, fusion or splitting events, short trajectory duration, motion blur, and heterogeneous short-range or long-range motion behaviors, require laborious and time-consuming manual tuning of empirical parameter choices [6–11]. Numerous solutions have been proposed to tackle these limitations and have been rigorously tested under various experimental conditions in open tracking challenges [12, 13]. However, algorithmic solutions are typically specific to a particular biological scenario. The recent emergence of data-driven deep learning (DL) techniques holds great promise for fully automatic end-to-end processing of time-lapse movies and extraction of biologically meaningful information [8–10, 14–18].

Deep learning, a subset of machine learning, uses multilayer artificial neural networks to autonomously discover feature representations from input data and then apply the learned patterns for prediction. Rapid advances in the development of Graphics Processing Units (GPUs) and learning algorithms have led to the unprecedented success of DL methods in biomedical image segmentation and classification tasks [14, 16]. Moreover, DL has been successfully applied to each phase of the conventional SPT analysis pipeline. Specific examples include (1) automatic classification of fluorescent spots at the surface of live cells as single or multiple quantum dots (QDs) via convolutional neural network (CNN) [19], (2) recognition of the 3D orientation of nanoparticles transported in living cells based on defocused image pattern analysis via CNN [20], (3) single cell and single molecule trajectory reconstruction via graph neural network (GNN), and (4) temporal annotation of diffusive state switching along single particle trajectories via CNN [17].

In this chapter, detailed steps for implementing two open-source DL methods are provided. First, diffraction-limited single QD spots in time-lapse image series are localized with subpixel accuracy using DeepTrack, a Python software package based on U-Net, a widely used type of CNN. Second, single particle trajectories and motion parameters are directly from raw time-lapse image series using SPT-net, a comprehensive end-to-end Transfomer-based DL architecture. Additionally, detailed instructions are provided to troubleshoot common errors encountered when installing and implementing these DL tools.

2 Materials

2.1 Workstation and Dataset

1. Windows or Linux workstation with Intel i7 or comparable CPU and at least 128 GB RAM.

2. CUDA (Compute Unified Device Architecture)-compatible NVIDIA GPU (RTX A6000 used in this work).
3. Dataset 1: 128 × 128 pixel^2 images of quantum dots diffusing on the live cell surface. Dataset 2: 30 frames of 64 × 64 pixel^2 images of QDs diffusing in the cytoplasm of a live cell.

2.2 Software for Data Analysis

1. ImageJ 1.54 g.
2. Anaconda or miniconda distribution (Python ≥3.9).
3. CUDA toolkit ≥11.8 and a compatible CUDA Deep Neural Network library (cuDNN).
4. Git 2.41.0 or later release.
5. DeepTrack 2.0.0 (https://github.com/DeepTrackAI/DeepTrack2).
6. SPTnet 1.0 (https://github.com/HuanglabPurdue/SPTnet).
7. MATLAB 2021b or later version.

3 Methods

3.1 DeepTrack Implementation

DeepTrack is a comprehensive, modular Python library for generating and analyzing single particle image data below the diffraction limit [8, 9]. The typical DeepTrack workflow consists of (a) synthetic training data preparation, which uses the image generation module to generate synthetic images of a single or multiple particles in the field of view (FOVs) based on a known microscope point spread function (PSF) corrupted by Gaussian and Poisson noise and (b) training and evaluation of a DL model, which uses one of the standard DL architectures including CNN, U-Net, generative adversarial network (GAN), and GNN. The trained DL model is subsequently applied to a series of experimental images to predict particle position with sub-pixel accuracy. In this section, the protocol for implementing the standard 2D U-Net model [21] to predict quantum dot centroid positions in a series of 2D images is described (Fig. 1). The 2D U-Net receives an image that may or may not contain quantum dots as an input and outputs a probability map of quantum dot location. For this segmentation task, the U-Net uses a contracting path (encoder), which captures image features and reduces spatial resolution through a series of convolutions, activations, and max pooling operations, and an expanding path (decoder), which up samples the feature maps to produce the final probability map.

1. Launch Anaconda command prompt. You can access it by clicking on the Start Menu and typing "Anaconda Prompt"

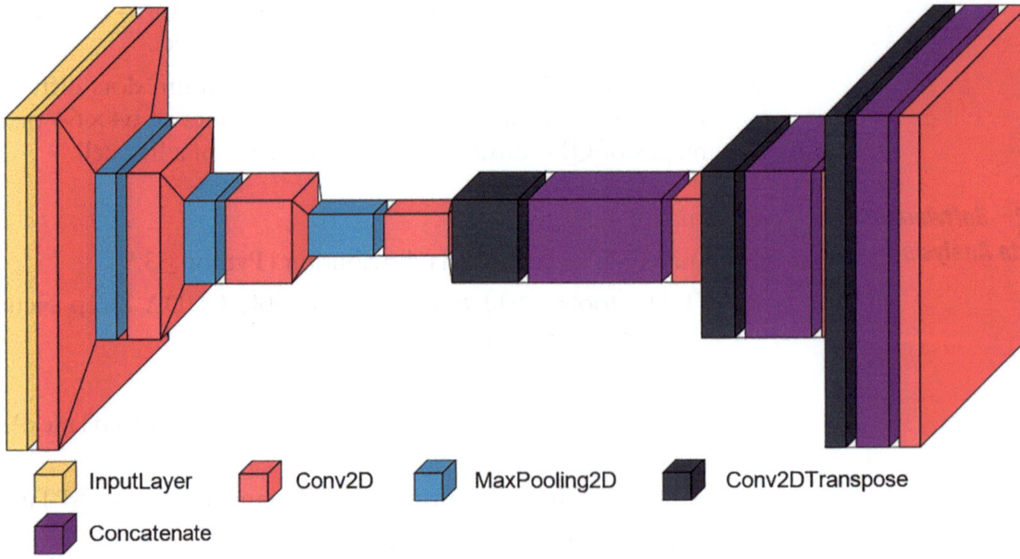

Fig. 1 Visual representation of the 2D U-Net architecture used for DeepTrack quantum dot detection. The number of features is doubled at each downsampling step in the contracting path (16, 32, and 64 features). The plot was generated using visualkeras (https://github.com/paulgavrikov/visualkeras)

in the search bar. Click on the Anaconda Prompt app to launch a new command prompt window (*see* **Note 1**).

2. Set up a new virtual environment for DeepTrack with the Python version of at least 3.9 by typing the following command into the command prompt:

```
conda create -n deeptrack_unet python=3.9 -y
```

3. Activate the environment and install the required libraries using pip (*see* **Note 2**):

```
conda activate deeptrack_unet
pip install notebook
pip install deeptrack
pip install deeplay
```

4. Launch Jupyter Notebook by typing "jupyter notebook" into the command prompt. Open the notebook DTGS131, which can be downloaded from the /tutorials/1-getting-started subfolder of the DeepTrack GitHub repository.

5. Execute code cells in Sects. 1–7 of the notebook to create synthetic images of point particles appearing as diffraction-limited blobs on noisy background. Carefully define the parameters of the PointParticle class (intensity and random xy position), Fluorescence class (i.e., the optical system), and

two sources of noise, a background offset and a Poisson noise with a random SNR. To match the properties of your optical setup, such as the numerical aperture (NA) of the objective, effective magnification, and quantum dot maximum emission wavelength, adjust the following code cell in Sect. 2 of the notebook (*see* **Note 3**):

```
fluorescence_microscope = dt.Fluorescence(NA=
0.7, resolution=1e-6, magnification=10, wave-
length=680e-9, output_region=(0, 0, 256, 256))
```

6. Configure the 2D U-Net model by defining its hyperparameters in Sect. 8 of the notebook. The default number of features computed at each 2D convolution layer is 16, 32, and 64. For the binary image classification task, binary cross-entropy (BCE) loss function (i.e., a measure of error between predicted output and actual target value) and the Adam optimizer (i.e., an algorithm that updates model parameters to minimize the loss function) with a learning rate (i.e., optimization step size) of 1e-4 are recommended starting points (*see* **Note 4**).
7. Train the model and visualize its performance by executing code cells in Sects. 9 and 10 of the notebook. Begin train ing with a small number of epochs, 5 or 10, (i.e., full cycle through the training data), as this is the most time-consuming step in the process. The default batch size is 4, equivalent to four training images processed in one training step of the epoch.
8. Once model training is complete and you are satisfied with the model performance (*see* **Note 4**), proceed to predict experimentally acquired image series in a new code cell of the notebook:

```
import tifffile
tif_path = 'qdot001.tif'
stack = tifffile.imread(tif_path)
image = stack[0,:,:]
### convert the image to the format (batch, channels,
width, height)
image = np.expand_dims(image, axis=-1)
image = np.expand_dims(image, axis=0)
image_transposed = np.transpose(image, (0, 3, 2,
1))
image_transposed = image_transposed.astype(np.
float32)
predicted_probabilities = model.predict(image_
transposed).sigmoid()
plt.figure(figsize=(10, 5))
plt.subplot(1,2,1)
```

```
plt.imshow(image_transposed[ 0, 0,...].squeeze(),
cmap="gray")
plt.axis('off')
plt.subplot(1,2,2)
plt.imshow(predicted_probabilities[0, 0,...])
plt.axis('off');
plt.tight_layout()
```

9. Figure 2 shows an example output of input image and output probability map for a single frame of the input time series. To process the entire time series, simply setup a for loop to predict one frame at a time.
10. Save the trained model using torch library, save the Jupyter notebook progress, and close it. Use Ctrl+C shortcut to end the Jupyter notebook instance and deactivate the virtual environment using "conda deactivate" command.

3.2 SPTnet Implementation

Ideally, the SPT deep learning framework should be a comprehensive solution that eliminates sequential processing from localization to trajectory linking to motion analysis in favor of a fully automatic, parameter-free, end-to-end pipeline. Recently developed SPTnet is an example of such a solution [11], as it employs a state-of-the-art combination of a CNN backbone, a dual Transformer proven effective in large language models (LLMs) [22], and a multitask multilayer perceptron (MLP) (Fig. 3). The CNN backbone is based on a popular spatiotemporal 3D ResNet neural network [23] and receives a video input of 30 frames of a 64 × 64 pixel^2 FOV (30 × 64 × 64) to extract initial features. The framewise features and whole video features are then processed by the spatial Transformer and the temporal Transformer, respectively, to yield temporal and fused (spatial-temporal)

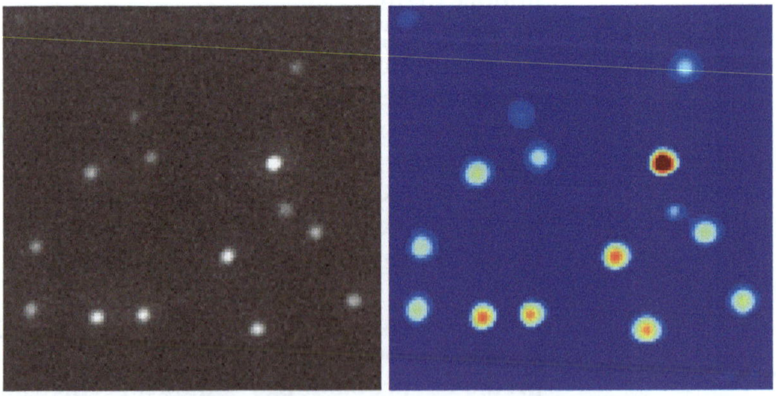

Fig. 2 DeepTrack detection of quantum dots in a single frame of the time series. Left: a 128 × 128 pixel^2 (14.96 × 14.96 μm^2) image of quantum dots appearing as diffraction-limited blob on dark background is shown. Right: a probability map of detecting individual quantum dots via DeepTrack is shown

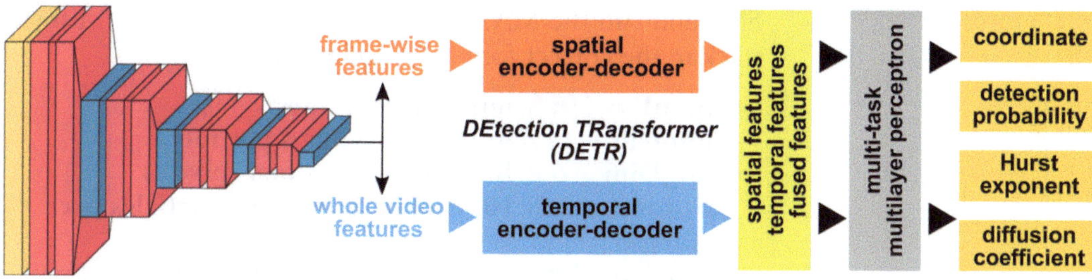

Fig. 3 Visualization of the end-to-end Transformer-based architecture of SPTnet

output. This output is then analyzed by the multitask MLP that estimates particle detection probability, its centroid position, and the diffusion coefficient, as well as the Hurst Exponent associated with the reconstructed trajectory.

1. Launch Anaconda command prompt and create a new virtual environment for SPTnet with the Python version of 3.9 by typing the following command into the command prompt:

```
conda create -n deepspt python=3.9 -y
```

2. Activate the virtual environment, clone the SPTnet GitHub repository, and install the required libraries (*see* **Note 5**):

```
conda activate deespt
git clone https://github.com/HuanglabPurdue/
SPTnet.git
conda install scipy h5py tqdm matplotlib pandas
IPython
pip install opencv-python
pip install torch==2.5.1 torchvision==0.20.1
torchaudio==2.5.1 --index-url https://download.
pytorch.org/whl/cu118
pip install positional-econdings[pytorch]
pip install visdom
```

3. The cloned repository contains the pretrained model "Pretrained_model_for_Ti2microscope" in the "Trained models" subfolder, which you will use for running inference of your experimental dataset. Before inference, navigate to the cloned directory using cd command (e.g., cd C:\Users\kovtuno\SPTnet).

4. If your time series are already stored as the MAT array, run inference in the command line:

```
python SPTnet_inference.py
```

   When prompted select the pretrained model for inference

5. If not, split your time series into 64 × 64 × 30 substacks (chunks) in ImageJ (*see* **Note 6**), save these substacks as individual tiff

files, and run the MATLAB script below to save each tiff sub-stack in the correct format for inference:

```
numF = 30; % number of frames
numvid = 1; % number of videos per file
imgDim = 64; % image dimensions (pixels)
p_num_max = 10; % max number of particles per
video
timelapsedata = single(zeros(imgDim, imgDim,
numF, numvid));
bglabel = zeros(1, numvid, 'single');
Perlinbglabel = zeros(1, numvid, 'single');
traceposition = cell(numvid, p_num_max);
Hlabel = cell(numvid, p_num_max);
Clabel = cell(numvid, p_num_max);
photonlabel = cell(numvid, p_num_max);
moleculeid = cell(numvid, p_num_max);
filename = qdot001.tif';
tiffObj = Tiff(filename, 'r');
numImg = length(imfinfo(filename));
imageStack = [];
  for k = 1: numImg
  tiffObj.setDirectory(k);
  imgData = tiffObj.read();
  if k == 1
    imageStack = zeros([size(imgData), numImg],
class(imgData));
  end
  imageStack(:,:,k) = imgData;
end
tiffObj.close();
timelapsedata = imageStack;
saveFileName = ['qdot_movie_', num2str(1)];
save(saveFileName, 'timelapsedata', 'Hlabel', 'Cla-
bel', 'photonlabel',
  'bglabel', 'traceposition', 'moleculeid', 'Perlin-
bglabel', '-v7.3');
```

6. When inference is completed, the output MAT file will be stored in the "inference_results" subfolder of the selected training model directory. This file will contain the reconstructed trajectories (Fig. 4), detection probabilities, Hurst exponents, and generalized diffusion coefficients.
7. Deactivate the virtual environment using "conda deactivate" command and close the Anaconda Prompt window.

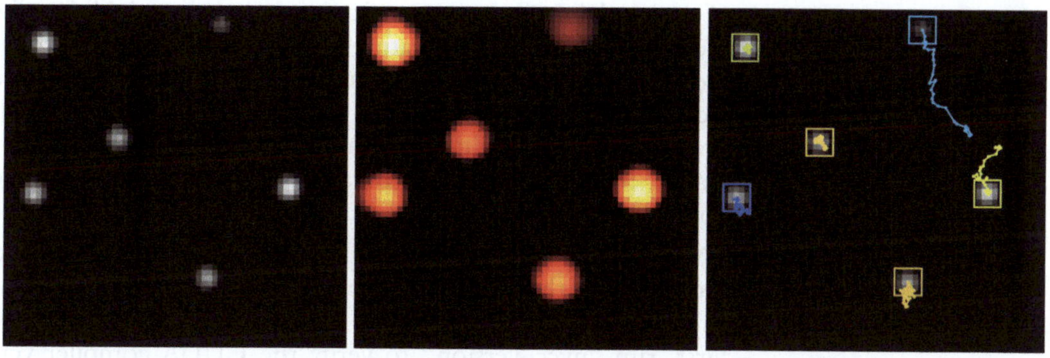

Fig. 4 SPTnet reconstruction and analysis of quantum dot trajectories in raw time series. Left: a 64 × 64 pixel^2 (4.16 × 4.16 μm^2) image of quantum dots appearing as diffraction-limited blob on dark background is shown. Center: a probability map of detecting individual quantum dots via SPTnet is shown. Right: reconstructed trajectories are displayed for each quantum dot present in the field of view

4 Notes

1. Verify your Anaconda or miniconda installation by running "conda–version" in the terminal window. Periodically, update conda by running "conda update conda" in the terminal window. To view the existing virtual environments in order to avoid conflict during installation, run either "conda env list" or "conda info–envs."
2. Dependency conflict is a common source of error when executing the code. It occurs when two or more libraries require different versions of the same package (e.g., numpy). To avoid dependency conflicts, follow the steps in the same order as described in this protocol. If the conflict occurs, check the requirements.txt file in the original GitHub repository for the correct library version and install the correct version via conda or pip. For instance, the DeepTrack configuration used in the development of this protocol included deeplay 0.1.2, deeptrack 2.0.0, numpy 1.26.4, and pint 0.19.2.
3. Adjustments of NA and total effective magnification may need to be accompanied by adjustments of the particle intensity and the output region size. It is recommended to use the default parameters for the initial training run, as detection of individual quantum dots in images acquired using a high NA objective and high magnification will be satisfactory.
4. Increasing the number of features extracted at each convolution and decreasing the learning rate may improve the performance

of the model at the expense of training duration. Additionally, the increased number of features can lead to overfitting, when the model does not generalize to new data. The default metric (accessed through the torchmetrics library) is F1 score, which is a good starting point for balanced data and general performance. BinaryAccuracy and BinaryPrecision (high cost of false positives) are examples of other useful metrics to consider.

5. SPTnet relies on the torch library to perform fast GPU-based processing of the input image series. Therefore, it is important to verify that the GPU version of torch is correctly installed. First, run "nvcc–version" to verify the CUDA compiler version installed. Next, activate the SPTnet virtual environment and launch python by typing "python" in the command line. Verify torch CUDA compatibility by running "import torch" followed by "torch.cuda.is_available()" and "torch.cuda.device_count()." If CUDA is not available, uninstall torch and reinstall its GPU version. NVIDIA driver update and CUDA reinstallation may also be needed.
6. Image stack can be easily split into 64 × 64 × 30 stacks in ImageJ using either a combination of Edit > Selection > Specify and Image > Stacks > Tools > Make Substack or a custom macro. The current SPTnet model is optimized for analyzing smaller data blocks of 64 × 64 pixel^2 over 30 frames. To process larger datasets, it is necessary to divide videos into smaller overlapping blocks (shifting 32 pixels along xy dimensions), each 30 frames long. The results are reassembled after each subblock is segmented via SPTnet.

Acknowledgments

Image analysis was performed in part through the use of the Vanderbilt Cell Imaging Shared Resource (supported by NIH grants CA68485, DK20593, DK58404, DK59637, and EY08126).

References

1. Crocker JC, Grier DG (1996) Methods of digital video microscopy for colloidal studies. J Colloid Interface Sci 179(1):298–310. https://doi.org/10.1006/jcis.1996.0217
2. Nguyen TD, Chen YI, Chen LH, Yeh HC (2023) Recent advances in single-molecule tracking and imaging techniques. Annu Rev Anal Chem (Palo Alto Calif) 16(1):253–284. https://doi.org/10.1146/annurev-anchem-091922-073057
3. Simon F, Weiss LE, van Teeffelen S (2024) A guide to single-particle tracking. Nat Rev Methods Primers 4(1):66. https://doi.org/10.1038/s43586-024-00341-3
4. Rosenthal SJ et al (2011) Biocompatible quantum dots for biological applications. Chem Biol 18(1):10–24. https://doi.org/10.1016/j.chembiol.2010.11.013
5. Kovtun O et al (2018) Single quantum dot tracking illuminates neuroscience at the nanoscale.

Chem Phys Lett 706:741–752. https://doi.org/10.1016/j.cplett.2018.06.019
6. Kovtun O, Thal LB, Josephs T, Rosenthal SJ (2020) Quantitative analysis of single quantum dot trajectories. In: Fontes A, Santos BS (Eds.) Eds. Quantum dots: applications in biology. Springer, US, New York, NY, pp 109–123
7. Kovtun O, Rosenthal SJ (2022) Methodological nuances of measuring membrane protein nanoscopic organization: a case of dopamine transporter. J Electrochem Soc 169(4):047505. https://doi.org/10.1149/1945-7111/ac60ef
8. Helgadottir S, Argun A, Volpe G (2019) Digital video microscopy enhanced by deep learning. Optica 6(4):506–513. https://doi.org/10.1364/OPTICA.6.000506
9. Midtvedt B et al (2021) Quantitative digital microscopy with deep learning. Appl Phys Rev 8(1):011310. https://doi.org/10.1063/5.0034891
10. Pineda J et al (2023) Geometric deep learning reveals the spatiotemporal features of microscopic motion. Nat Mach Intell 5(1):71–82. https://doi.org/10.1038/s42256-022-00595-0
11. Bi C et al (2025) SPTnet: a deep learning framework for end-to-end single-particle tracking and motion dynamics analysis. bioRxiv 2025.02.04.636521. https://doi.org/10.1101/2025.02.04.636521
12. Ulman V et al (2017) An objective comparison of cell-tracking algorithms. Nat Methods 14(12):1141–1152. https://doi.org/10.1038/nmeth.4473
13. Muñoz-Gil G et al (2021) Objective comparison of methods to decode anomalous diffusion. Nat Commun 12(1):6253. https://doi.org/10.1038/s41467-021-26320-w
14. Seo H et al (2020) Machine learning techniques for biomedical image segmentation: an overview of technical aspects and introduction to state-of-art applications. Med Phys 47(5):e148–e167. https://doi.org/10.1002/mp.13649
15. Yao Y et al (2020) Deep-learning method for data association in particle tracking. Bioinformatics 36(19):4935–4941. https://doi.org/10.1093/bioinformatics/btaa597
16. Liu Z et al (2021) A survey on applications of deep learning in microscopy image analysis. Comput Biol Med 134:104523. https://doi.org/10.1016/j.compbiomed.2021.104523
17. Kæstel-Hansen J et al (2025) Deep learning-assisted analysis of single-particle tracking for automated correlation between diffusion and function. Nat Methods 22: 1091–1100. https://doi.org/10.1038/s41592-025-02665-8
18. Kovtun O (2025) Machine learning classification of 3D intracellular trafficking using custom and imaris-derived motion features. Receptors 4(1):6. https://doi.org/10.3390/receptors4010006
19. Wang Q et al (2021) Deep-learning-assisted single-molecule tracking on a live cell membrane. Anal Chem 93(25):8810–8816. https://doi.org/10.1021/acs.analchem.1c00547
20. Song D et al (2024) Deep learning-assisted automated multidimensional single particle tracking in living cells. Nano Lett 24(10):3082–3088. https://doi.org/10.1021/acs.nanolett.3c04870
21. Ronneberger O, Fischer P, Brox T (2015) U-net: convolutional networks for biomedical image segmentation. In: Navab N, Hornegger J, Wells WM and Frangi AF (eds) Medical image computing and computer-assisted intervention—MICCAI 2015. Cham, Springer International Publishing, pp 157
22. Carion N et al (2020) End-to-end object detection with transformers. In: In: Vedaldi A, Bischof H, Brox T and Frahm JM (eds) Computer Vision – ECCV 2020. Lecture Notes in Computer Science, 12346: 13-229. Springer, Cham. https://doi.org/10.1007/978-3-030-58452-8_13
23. Kataoka H, Wakamiya T, Hara K, Satoh Y (2020) Would mega-scale datasets further enhance spatiotemporal 3D CNNs? arXiv:2004.04968. https://doi.org/10.48550/arXiv.2004.04968

Chapter 7

Machine Learning for Optimizing the Optical Properties of Quantum Dots

Cintia Ellen Giarola, Fernanda Sumika Hojo de Souza and Marco Antônio Schiavon

Abstract

The application of machine learning (ML) to the synthesis of quantum dots (QDs) represents a significant innovation in material optimization. Due to the quantum confinement effect, the optical properties of QDs can be tuned by controlling the size of the material during synthesis, but this traditionally requires a long time of experimental synthesis practice. ML is an efficient approach that analyzes databases of synthesis variables to predict final properties, thereby reducing the experimental effort required. Supervised and unsupervised models play complementary roles, in this context, helping to predict properties or identify patterns in unlabeled data. In this chapter, the main steps that include the construction of structured databases, feature engineering, and model training for accurate predictions are presented, highlighting the potential of ML in developing QDs with specific optical properties.

Keywords Machine learning, Quantum dots, Materials optimization, Quantum dot synthesis, Optical properties

1 Introduction

The use of machine learning (ML) to predict the synthesis patterns of quantum dots (QDs) represents a significant advance in the optimization of these materials and is, therefore, being explored in various studies [1–3]. Due to the quantum confinement effect, the optical properties of QDs, such as light absorption and emission, could be adjusted by controlling the size of the material during its synthesis [4, 5]. However, controlling these properties requires a laborious process of trial and error to identify the ideal conditions. ML provides an efficient solution by analyzing databases of synthesis variables and predicting the final properties of the material, thereby reducing the number of experimental trials required.

ML is an area of artificial intelligence that can identify patterns in data and use them to make predictions or decisions under uncertainty.

Goreti Pereira et al. (eds.), *Quantum Dots*, Methods in Molecular Biology, vol. 3025,
https://doi.org/10.1007/978-1-0716-5210-7_7,

ML models improve their performance incrementally as they gain experience with the data, without having to be explicitly programmed for each task [6]. The main types of ML models are supervised and unsupervised models, which differ in the way data are used to learn [7].

Supervised models work with labeled data sets, i.e., they have known input variables (such as temperature, synthesis time, and proportion of reagents) associated with the desired output (such as emission wavelength or QD size). These models learn to relate inputs and outputs, making them capable of predicting material behavior under new conditions. In contrast, unsupervised models deal with data without predefined labels and are useful for identifying hidden patterns or grouping similar conditions without necessarily predicting a specific output [7].

ML has already been used to study the influence of the main synthesis variables on the light emission of CdTe QDs using a database of articles published in the literature. This study revealed a strong dependence of the emission wavelength on the reaction time, surface ligand, and precursor concentrations, making it possible to adjust these variables to optimize the synthesis conditions and obtain emissions at specific wavelengths [3].

In another study, ML was used to predict the ideal conditions for synthesizing luminescent thin films of AgZnInS QDs based on experimental data obtained by the authors themselves. The predicted results were experimentally confirmed, demonstrating the potential of ML to facilitate, and improve the synthesis of advanced materials [2].

The source code of the ML model for both applications can be found in the published articles [2, 3]. Following this approach, ML stands out as a method for analyzing databases that relate synthesis variables to the final properties of materials. By identifying patterns in these data, ML speeds the search for optimized conditions to obtain the desired properties.

This methodology presents an introductory approach using the Python programming language and PyCaret library to build a supervised ML model aimed at optimizing the optical properties of QDs. This chapter discusses fundamental steps, such as creating a structured database, selecting and preparing the most relevant variables (feature selection and engineering), and training an ML model to make accurate predictions of the material's synthesis parameters. The following sections describe each step in detail and demonstrate how the use of ML can improve the design of QDs with specific properties.

2 Materials

2.1 Development Environment

1. Python: version 3.6 or higher installed in the development environment, either locally or in a cloud service.
2. Integrated Development Environment (IDE): Google Colab (*see* **Note 1**).

2.2 Libraries and Packages

1. Pandas: library for data manipulation and database organization.
2. NumPy: library for numerical calculations, matrix operations, and handling large volumes of data.
3. Scikit-learn: this library is used to implement machine learning algorithms, such as regression, classification, and hyperparameter optimization.
4. PyCaret: Automated Machine Learning (AutoML) library for rapid implementation and comparison of multiple ML models, cross-validation, and model optimization.
5. GridSearchCV: used for hyperparameter optimization during ML model training.
6. Matplotlib/Seaborn: the libraries for data and result visualization.

2.3 Training Data

1. Spreadsheet Software: Microsoft Excel or Google Sheets, used to collect, organize, and structure data before use by machine learning models. These software tools allow the recording of input data (features) and output data (labels) in table format, which can be exported as CSV files or directly connected to Python code via libraries such as Pandas.

3 Methods

3.1 Data Collection and Database Construction

1. Identify and collect the necessary data for analysis, ensuring that the data include both input and output variables (*see* **Note 2**).
2. Obtain data manually from published articles, theses, and dissertations, or use your own properly documented experimental results.
3. Use tools like WebPlotDigitizer to extract numerical data from graphs, if applicable.
4. Organize the collected data into a structured database with instances (rows) and features/attributes (columns) (*see* **Note 3**).
5. Export data to CSV format for processing in Python environment. For educational purposes, we refer to this file "ML_data" but it can be modified, if necessary.
6. Copy the file from your computer into your Google Drive account. For educational purposes, we placed this file in a folder called "QDs ML." The proposed configuration can also be modified.

3.2 Importing Data into Programming Environment

1. When Google Colab is open, first, the PyCaret library package is installed using the following command:

```
!pip install pycaret
```

2. Import the libraries used throughout the code:

```
import pandas as pd
import seaborn as sns
import matplotlib.pyplot as plt
import numpy as np
```

3. Connect to the Google Drive folder where the file is located:

```
from google.colab import drive
import os
drive.mount('/content/drive')
os.chdir("/content/drive/MyDrive/QDs ML")
```

4. The CSV file is imported and assigned a name: (*see* **Note 4**)

```
dataset = pd.read_csv(' ML_data.csv')
```

3.3 Feature Engineering

1. Analyze the raw data to understand the distributions of the variables and apply transformations that may be interesting for the ML model (*see* **Note 5**).

2. Observe the data volume and identify columns with a high percentage of missing data to eliminate them (Fig. 1):

```
filled_by_column = (1-dataset.isna().mean())*100
plt.figure(figsize=(12, 8))
sns.barplot(x= filled_by_column.index, y= filled_
by_column.values, palette='viridis')
plt.title(' Data Completeness')
plt.xlabel('Columns')
plt.ylabel(' % filled')
plt.xticks(rotation=45)
plt.show()
```

3. Remove any remaining missing data using the following command:

```
dataset = dataset.dropna() (see Note 6).
```

4. Build the correlation matrix to understand the relationship between the variables (Fig. 2) using the following command (*see* **Note 7**):

```
def plot_corr(dataset, filename):
    corr = dataset.corr(method='spearman')
    sns.heatmap(corr,
        xticklabels=corr.columns,
        yticklabels=corr.columns, cbar=True,
square=True, annot=True, annot_kws={'size': 5},
cmap='coolwarm')
    strname = str(filename) + '.png'
    plt.savefig(strname, dpi=300, transparent=
False, facecolor='w', bbox_inches='tight')
plot_corr(dataset, 'fig1_3')
```

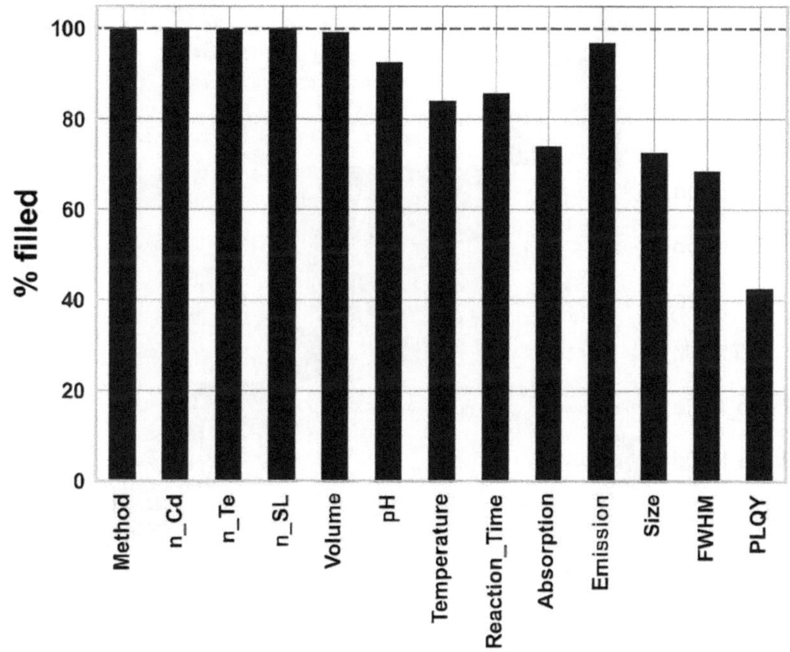

Fig. 1 Percentage of data for each variable in the dataset. Reproduced from [3] with permission from AIP Publishing

3.4 Building Machine Learning Model

1. First, the preprocessing environment is setup as follows:

```
s = setup(data=dataset, target='Emission')
```
(*see* **Note 8**)

2. Using setup, split the dataset into training and testing sets, for example, 85% for training and 15% for testing. However, this proportion should be varied and analyzed to determine which yields the best results depending on the dataset. It is recommended not to use a very high value for testing. Common splits are 80%–20% or 70%–30% for training and testing, respectively. Avoid test sizes below 5%, which may lead to inaccurate evaluation, and above 40%, which may result in insufficient training. For very large datasets, a smaller test set may be sufficient:

```
s = setup(data=dataset, target='Emission', train_size=0.85)
```

3. Analyze the model performance metrics:

```
best = compare_models()
```
(*see* **Note 9**)

4. Test different parameters of the setup command and check whether there is an improvement in the model performance metrics (*see* **Note 10**).

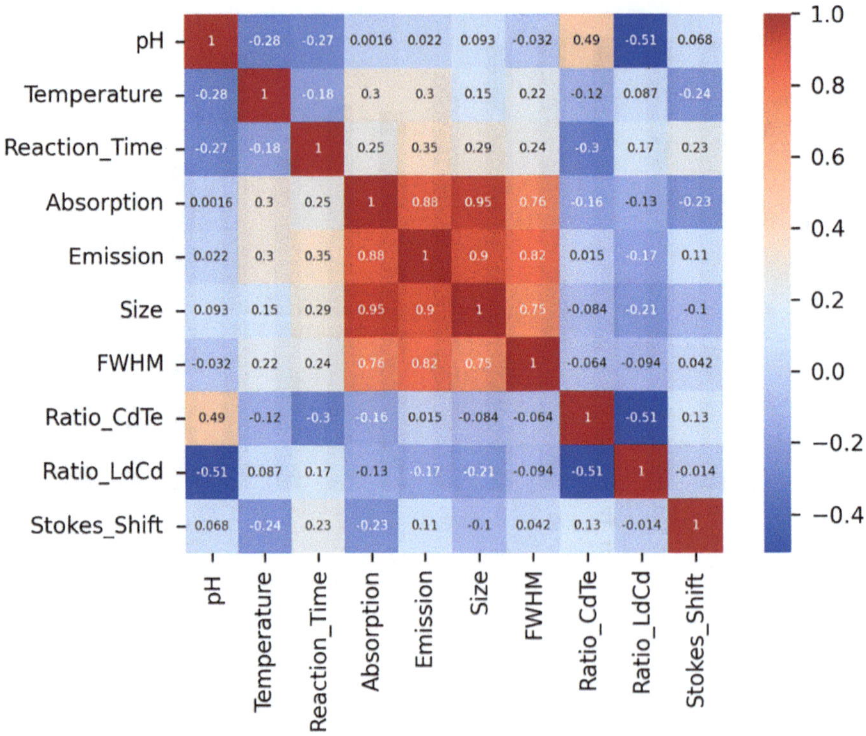

Fig. 2 Correlation matrix. Reproduced from [3] with permission from AIP Publishing

5. Experiments with different algorithms, such as decision trees, neural networks, and linear regression:

```
model = create_model('et') (see Note 11)
```

6. Optimize the ML model's hyperparameters using Grid Search.
7. Analyze the models using plot_model and select the model with the best results (*see* **Note 12**).
8. Plot different evaluation graphs, such as prediction error, residuals, and feature importance (Fig. 3) (*see* **Note 13**):

```
plot_model(model, plot='error')
plot_model(model, plot='residuals')
plot_model(model, plot='feature')
```

3.5 Optimized Synthesis Using ML Results

1. To evaluate the model's performance, use the predict_model function. This function makes predictions on the test data, which were previously separated and whose actual values are known as follows:

```
predict = predict_model(model)
```

2. To visualize predictions, use the following command:

```
print(predict.head())
```

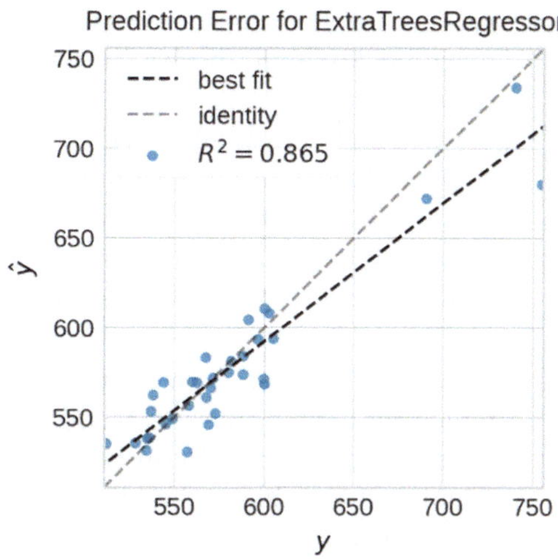

Fig. 3 Prediction Error Plot for ExtraTreesRegressor. Reproduced from [3] with permission from AIP Publishing

3. Prepare the dataset for predictions. It is important to ensure that this dataset is different from the one used for training. These new data represent inputs whose outcomes are not yet known. To achieve this, use an external dataset or manually create a dataset containing only the predictor variables. Use the following command:

```
new_data = pd.read_csv("new_ML_data.csv")
```

4. Remove the target variable, for example, "Emission":

```
new_data.drop('Emission', axis=1, inplace=True)
new_data.head()
```

5. Generate predictions on the dataset without the target variable to predict missing values:

```
predictions = predict_model(model, data=new_data)
```

6. To visualize the table with predicted values:

```
predictions.head()
```

7. The synthesis parameters predicted by the created model are now available. You can export these files back to a CSV file using the following command:

```
predictions.to_csv("predicted_values.csv",
index=False)
```

4 Notes

1. IDE (Integrated Development Environment) is a software that brings together all the necessary tools for programming, such as a code editor, debugger, and compiler, in one place.

In this methodology, we have described how to program machine learning models in Python using Google Colab, which is an IDE based on Jupyter Notebook and hosted on Google Cloud, allowing for online collaboration. Other popular IDEs for programming in Python include Jupyter Notebook, PyCharm, VSCode, and IDLE, each with its advantages and disadvantages. You can choose the one that best suits your needs, but be aware that the syntax of the programming lines may vary between them.

2. To apply ML models, it is necessary to provide a database in which the algorithm can analyze patterns and make decisions. This database should consist of input and output variables. The input variable is the information provided to the model so that it can make predictions or decisions. The output variable is the result that the model generates in response to the input provided. It is important to note that in the supervised ML model training stage, the instances must be associated with the output variables so that the algorithm can learn from this dataset. To study the optical properties of QDs, common input variables would be: molar quantity of the precursors, volume of the surface ligand and solvent(s), pH in the case of syntheses where it varies, reaction temperature, and reaction time. Typically output variables include: absorption and emission wavelength, as well as intensity (if relevant to the study), nanoparticle size, full width at half maximum (FWHM), photoluminescence quantum yield, and emission lifetime. These variables can be modified according to the study being conducted, but it is important that the input variables relate to the output variables.
3. The example columns include the reagent concentrations, synthesis temperatures, and emission wavelengths. The rows are the instances associated with the values in the columns. In the case of collecting data from the literature, it is recommended to identify the data of each row with the DOI of the article, from which the information was obtained and give credit to the author.
4. The command `pd.read_csv('ML_data.csv')` loads the file into Google Colab. Once it is loaded, it is stored in a variable that we call "dataset" here, but it can also be modified if desired. From this point on, we can use this variable to manipulate the data in the file.
5. Feature engineering is important to ensure that the ML model learns better from the data. The programming environment can be used to create, transform, and extract variables from the raw data to improve the model's performance. For example, the molar concentration of precursors can be

calculated using the molar quantity and volume of the collected solution. From this point on, the ratio between the precursors can be calculated to understand their proportion and its relationship to the final property of the material. In addition, the Stokes shift can be calculated by the difference between the emission and the absorption wavelength.

6. In the context of a database, missing data refer to entries that do not have value. In particular, if the database was built from articles in the literature, sometimes some variable is not reported, making it impossible to register them. The command `dataset.dropna()` removes all rows from the dataset that contain missing data from your database. The result is a new version of the dataset, which does not have entries with missing values. If a column has a very high percentage of missing data, such as QY in Fig. 1, it is necessary to remove this column from the database prior to this step because otherwise, the volume of data will decrease drastically, which can harm the performance of the algorithm. Another possibility is to impute missing data using techniques such as the mean, median, or mode (depending on the data distribution) to minimize the loss of information.
7. A correlation matrix is a table that shows the intensity and direction of correlations between pairs of variables in a dataset. Each element in the matrix ranges from −1 to 1, indicating the strength and nature of the relationship between the variables: a value of 1 represents a perfect positive correlation, −1 represents a perfect negative correlation, and 0 indicates no correlation. It is used to understand how different variables are interconnected. Through the correlation matrix, it is possible to select the most relevant variables. Highly correlated variables should be avoided because they can result in multicollinearity issues [8, 9].
8. The `setup` function initializes the training environment. Note that this function should be called before executing any other function. Only two parameters are mandatory: "data" and "target," where "target" is the variable (present in the "data" dataset) that you wish to predict. Other parameters are optional. However, here, we used the training split, where 85% of the dataset was used for model training and 15% was used for testing. This can be modified according to your data.
9. The performance metric is used to evaluate the effectiveness and quality of the proposed ML model. In this study, we evaluate R2 (R^2), which measures the proportion of variability in the response variable explained by the predictor variables. The closer the value is to 1, the better is the model.

10. The parameters of the "setup" directly influence the behavior of the model. Each parameter controls the important aspects of the analysis process. They should be configured according to the specific database being analyzed; thus, the authors should test them individually for their dataset. The Pycaret website [10] is recommended for better understanding each parameter and testing them.
11. In the command `create_model('et', round = 2)`, the 'et' should be replaced with the abbreviation of the model to be evaluated. The `compare_models()` command returns a table containing the models and performance metric. The first column indicates the abbreviation used. In this example, 'et' stands for "Extra Trees Regressor."
12. With `plot_model`, it is possible to plot graphs to analyze model performance. The residuals graph shows each point as a value predicted by the model; the x-axis is the value present in the dataset, while the y-axis is the value predicted by the model. The closer the points are to the light line, the better the model, as it can predict more accurately.
13. The residual graph reveals nonrandom patterns that could indicate the need for model or data adjustments. The error graph shows the distribution of predictive errors, which helps assess the model's accuracy and identify whether the model is underestimating or overestimating values. Finally, the feature importance graph highlights the variables that are most influential in making predictions, allowing for the selection of the most relevant variables and the elimination of variables that do not add value.

Acknowledgments

The authors would like to thank the financial support from (FAPEMIG) Fundação de Amparo à Pesquisa do Estado de Minas Gerais), CNPq (Conselho Nacional de Desenvolvimento Científico e Tecnológico), CAPES (Coordenação de Aperfeiçoamento de Pessoal de Nível Superior), and FINEP (Financiadora de Estudos e Projetos).

References

1. Xing C, Chen G, Zhu X, An J, Bao J, Wang X, Zhou X, Du X, Xu X (2024) Synthesis of carbon dots with predictable photoluminescence by the aid of machine learning. Nano Res 17:1984–1989. https://doi.org/10.1007/s12274-023-5893-6
2. Wei S, Luo X, Shao S, Zhang L (2024) Ionic liquid-assisted preparation of Ag–Zn–In–S quaternary quantum dot thin films and luminescence performance optimized by machine learning. Dyes Pigm 223:111970. https://doi.org/10.1016/j.dyepig.2024.111970

3. Fonseca AFV, Giarola CE, Carvalho TAS, Hojo De Souza FS, Schiavon MA (2023) Machine learning predicted emission of water-stable CdTe quantum dots. J Chem Phys 159:184705. https://doi.org/10.1063/5.0170957
4. Silva FO, Viol LC, Ferreira DL, Alves JLA, Schiavon MA (2010) O estado da arte da síntese de semicondutores nanocristalinos coloidais. Quím Nova 33:1933–1939. https://doi.org/10.1590/S0100-40422010000900021
5. García de Arquer FP, Talapin DV, Klimov VI, Arakawa Y, Bayer M, Sargent EH (2021) Semiconductor quantum dots: technological progress and future challenges. Science 373:eaaz8541. https://doi.org/10.1126/science.aaz8541
6. Schleder GR, Fazzio A (2021) Machine learning in physics, chemistry, and materials science: materials discovery and design. Rev Bras Ensino Fís 43:e20200407. https://doi.org/10.1590/1806-9126-RBEF-2020-0407
7. Baloglu O, Latifi SQ, Nazha A (2022) What is machine learning? Arch Dis Child Educ Pract Ed 107:386–388. https://doi.org/10.1136/archdischild-2020-319415
8. Schreiber-Gregory DN, Waller J, Smith T (2018) Ridge regression and multicollinearity: an in-depth review. Model Assist Stat Appl 13:359–365. https://doi.org/10.3233/MAS-180446
9. Kim JH (2019) Multicollinearity and misleading statistical results. Korean J Anesthesiol 72:558–569. https://doi.org/10.4097/kja.19087
10. PyCaret Documentation (2024) Other setup parameters. https://pycaret.gitbook.io/docs/get-started/preprocessing/other-setup-parameters

Check for updates

Chapter 8

Imaging and Reconstruction of Three-Dimensional Quantum Dot Trafficking in Live Cells by Spinning Disk Confocal Microscopy

Oleg Kovtun

Abstract

Protocols are provided for time-lapse tracking of three-dimensional quantum dot (QD) movement in live cells using spinning disk confocal microscopy. The workflow employs a combination of a commercially available spinning disk confocal system and open-source ImageJ plugins, as well as the popular Imaris 3D/4D image analysis software. Implementation of the protocol streamlines all phases of the experiment: volumetric live-cell imaging, subsequent QD localization in 3D with nanometer accuracy, and continuous trajectory reconstruction. Detailed instructions are described for preparing QD control samples, labeling live cells with QD conjugates, and defining the appropriate parameters for image acquisition and QD trajectory reconstruction.

Keywords Quantum dots, Single particle tracking, Three-dimensional tracking, Endocytic trafficking, Spinning disk confocal microscopy, Deconvolution, Time-lapse imaging, Drift correction

1 Introduction

The ability to access the precise location of intracellular cargo and monitor its endocytic fate in real time holds great promise in three key areas—tracking viral infection [1, 2], enhancing the efficiency of targeted intracellular delivery of therapeutics [3], and achieving control over cell surface protein recycling in order to modulate its signaling from intracellular compartments [4, 5]. Although ensemble biochemical and imaging techniques could detect intracellular cargo accumulation and capture snapshots of endocytic segregation, live-cell single particle imaging techniques provide an experimental avenue to identify intracellular cargo position with nanometer accuracy and quantitatively map dynamic networks involved in endocytic cargo routing [6–9]. Single particle methods are now widely available, as the past decade witnessed quantum leaps in technological advances

Goreti Pereira et al. (eds.), *Quantum Dots*, Methods in Molecular Biology, vol. 3025,
https://doi.org/10.1007/978-1-0716-5210-7_8,

of instrumentation, image analysis algorithms, and probe development [6, 8].

Quantum dots (QDs) have emerged as a useful and versatile probe for intracellular cargo tracking, especially when combined with rapid acquisition and high penetration depth of spinning disk confocal (SDC) and light-sheet microscopes [10–15]. High photon budget, excellent colloidal stability in complex intracellular environments, and diversity of available surface functionalization approaches make QDs uniquely suitable for analyzing the distribution, intracellular state, and diffusion dynamics of internalized cargo [15, 16]. Notable examples of QD use in this field include identifying important carrier design parameters for nanoparticle-mediated drug delivery [10], shedding light on stepwise viral cell entry and uncoating [14], uncovering the selectivity mechanisms of nuclear pore complex-dependent cargo translocation [17], revealing distinct fusion modes of synaptic vesicle exocytosis [18], characterizing long-range axonal transport between spatially segregated neuronal compartments [19], and measuring the kinetics of cell surface protein endocytosis [20]. There has also been substantial progress in our understanding of the impact of physicochemical properties (i.e., surface chemistry, size, shape) of QDs on their intracellular trafficking [15].

In this chapter, detailed protocols are described to implement time-lapse volumetric imaging of QDs in live cells on a commercial spinning disk confocal (SDC) microscope. SDC microscopy provides optical sectioning at fast rates and low photodamage, an ideal combination for long-term, live-cell imaging. The use of QDs with large photon budget, high signal-to-noise ratio, and excellent photostability under continuous excitation naturally complements SDC microscopy to enable observation of biological processes over extended time periods. First, protocols are provided to prepare two key control samples—immobile, coverslip-adsorbed QDs and non-targeted QDs loaded into the cytoplasm of live cells. These control samples are then used to optimize conditions, including imaging buffer composition, camera exposure time, laser power, and dimensions of volumetric acquisition. Then, detailed instructions are given for labeling of a plasma membrane lipid in live cells and analysis workflows that consist of three-dimensional QD localization with nanometer accuracy over time and subsequent QD position linking into continuous trajectories. Additionally, protocols are described for commonly used artifact removal techniques, such as iterative image restoration via 3D deconvolution and drift correction.

2 Materials

2.1 Control Samples (Immobilized QDs and Internalized QDs)

1. Streptavidin-conjugated QDs with 605 nm emission wavelength.
2. Qtracker™ Cell Labeling kit, 605 nm emission wavelength.

3. Low background phenol-red free medium: FluoroBrite™ DMEM or Leibovitz's L-15 medium.
4. Human embryonic kidney cell line HEK-293 (ATCC CRL-1573).
5. Complete growth medium: Eagle's Minimum Essential Medium (EMEM) and 10% fetal bovine serum (FBS).
6. Imaging buffer. Option A: Leibovitz's L15 supplemented with 2% glucose, 2% dialyzed FBS, and 75 mM HEPES (4-(2-hydroxyethyl)-1-piperazineethanesulfonic acid). Option B: FluoroBrite DMEM supplemented with 2% dialyzed FBS and 75 mM HEPES (*see* **Note 1**). If substantial QD photobleaching is observed, it may be necessary to include a reducing agent (e.g., 5–20 mM dithiothreitol) or an oxygen scavenger (1–4 μmol/L glucose oxidase + 0.1–0.2 μmol/L catalase enzymatic system).
7. 35-mm No. 1.5 or 1.5H glass-bottom dish with 14 mm glass diameter (MatTek) or 4-well No. 1.5H glass-bottom chambered coverslip (Ibidi).

2.2 Live Cell Membrane Labeling of GM1 Gangliosides

1. Biotin-CTxB: biotinylated B subunit of cholera toxin.
2. Streptavidin-conjugated QD, emission wavelength 605 nm.
3. FluoroBrite DMEM or Leibovitz's L-15 supplemented with 2% dialyzed FBS.
4. Phenol red-free medium supplemented with 2% dialyzed FBS.
5. Optional: NucBlue™ Live ReadyProbes™ Reagent (Hoechst 33342).
6. Optional: QSY™ 21 extracellular QD fluorescence quencher.
7. Human embryonic kidney cell line HEK-293 (ATCC CRL-1573).
8. Complete growth medium: EMEM and 10% FBS.
9. 35-mm No. 1.5 or 1.5 H glass-bottom dish with 14 mm glass diameter (MatTek) or 4-well No. 1.5 H glass-bottom chambered coverslip (Ibidi).

2.3 Microscope Configuration

1. Nikon Ti2 inverted microscope equipped with the stage-top Tokai Hit incubator, Yokogawa CSU-W1 spinning disk unit, super-resolution by optical pixel reassignment (SoRa) disk, a magnification changer, Hamamatsu ORCA-Fusion BT sCMOS camera, 60× NA 1.42 Plan Apochromat λD objective, high-speed piezo Z stage, an automated and encoded X/Y stage, and a four-line solid-state laser launch (100 mW 405 nm, 100 mW 488 nm, 100 mW 561 nm, and 75 mW 640 nm).
2. Semrock Di01-T405/488/568/647-13×15×0.5 dichroic quad-beamsplitter.

3. 605/52 nm and 455/50 nm emission filters for QD and NucBlue detection, respectively.

2.4 Software for Data Analysis

1. ImageJ with TrackMate and Fast4DReg plugins installed.
2. NIS-Elements Analysis AR 5.42.
3. Imaris 10.2 (Bitplane).

3 Methods

3.1 Immobilized QD Preparation and Imaging

In this section, the protocol for preparing the immobilized QD control sample is described. The aim is to prepare a sample of QDs adsorbed to the glass coverslip at a density suitable for single particle detection (<1 QD/μm^2). Sparse QD density will allow the user to (1) assess the photoluminescence (PL) performance of QDs in the prepared imaging buffer, (2) examine the QD point spread function (PSF) for aberrations, and (3) determine the baseline velocity and diffusion rate of the immobile particles during the analysis stage.

1. Serially dilute the stock QD suspension (1 μmol/L) in the freshly prepared imaging buffer (*see* **Note 2**) to the final concentration of 1–2 pmol/L (10^6 dilution). Vortex at 2500 rpm for 30 s after each dilution step.
2. Pipet 300 μL of the resulting QD suspension to the glass coverslip center of the MatTek dish or 150 μL to a well of the chambered coverslip, respectively.
3. Incubate for 1 min at room temperature (RT; approximately 20°C) and gently aspirate the QD suspension using pipette.
4. Place the sample on the microscope stage with the lid off. It is recommended to turn on the stage-top incubator and image under the same conditions used during live-cell imaging.
5. Choose the appropriate optical configuration (OC) and locate the sample using the oculars. The standard TRITC OC works well for QD605. The focal plane of QDs adsorbed to the coverslip should be slightly above the *z* position at which the built-in Perfect Focus System (PFS) detects the refractive index mismatch at the coverslip interface.
6. Choose the standard W1 confocal mode, switch to the spinning disk OC, and set the camera exposure time and laser excitation power (*see* **Note 3**). Next, carefully add a fresh imaging buffer (2 mL for the MatTek dish and 700 μL for the four-well chambered coverslip) to the side wall (*see* **Note 4**).
7. Select the appropriate camera ROI (e.g., 512 × 512 pixels) and define your time-lapse duration. A good starting point is to

set the interval to no delay (continuous acquisition) and the number of loops to 50.

8. Next, define your *z*-stack acquisition parameters (Fig. 1). A good starting point is a 10 µm range at the recommended z-step size of 0.2 µm for the 60× objective, which is equivalent to 50 xy planes per *z*-stack. Select *Close Active Shutter During Z Movement* and choose *Triggered Piezo Z* from the Z Device drop-down list (*see* **Note 5**).
9. Click "Run Now" in the ND Acquisition panel to begin acquisition.

3.2 Internalized QD Preparation and Live-Cell Imaging

In this section, the protocol for intracellular loading of live cells with Qtracker QDs is described. The Qtracker probes are based on the standard core-shell QD encapsulated in an amphiphilic polymer. The Qtracker cell labeling kit also includes Component B, a proprietary cell-targeting peptide to deliver QDs into the cytoplasm of live cells. The goal of using Qtracker probes is to establish detection of QDs in the intracellular environment and measure motion parameters (e.g., velocity, diffusion rate) for nonimmobilized, nontargeted QDs.

1. Seed the cells at the density of 2–3 × 10^4 cells/cm^2 24 h prior to imaging and place the glass-bottom dish in the 37°C, 5% CO_2 incubator.
2. To prepare a 10 nM Qtracker labeling solution, premix 1 µL each of Qtracker Component A (2 µM stock concentration) and Component B in a 1.5 mL microcentrifuge tube. Incubate for 5 min at RT. Add 0.2 mL of fresh complete growth medium to the tube and vortex for 30 s.

Fig. 1 Z-stack acquisition settings in NIS-Elements

3. To achieve single-particle density in cytoplasm, add enough of the 10 nM Qtracker labeling solution to achieve the final concentration of 2–20 pmol/L (*see* **Note 6**).
4. OPTIONAL: add one to two drops of the NucBlue reagent to label the cell nucleus.
5. Incubate at 37°C for 15–30 min. Wash the cells with a warm imaging buffer several times and place the glass-bottom dish on the microscope stage.
6. Locate the sample, set the appropriate acquisition parameters, and begin acquisition as described in Steps 6–9 of Sect. 3.1

3.3 GM1 Ganglioside Labeling and Live-Cell Imaging

In this section, directions are provided to label the cell surface ganglioside GM1 with biotinylated CTxB and streptavidin-conjugated QDs in a two-step approach. GM1 has been shown to be involved in cholesterol-rich lipid raft-mediated endocytosis and undergoes marked endocytosis in live cells under basal conditions at 37°C, thus, serving as a reliable endocytic marker [21].

1. Seed the cells at the density of $2–3 \times 10^4$ cells/cm^2 at least 24 h prior to imaging and place the glass-bottom dish in the 37°C, 5% CO_2 incubator.
2. Ascertain that the cells express ganglioside GM1 endogenously through literature search (*see* **Note** 7). Incubate the cells for 10 min with 20–200 nmol/L biotinylated CTxB, which can be added directly into the cell chamber/dish with complete growth medium. OPTIONAL: add a drop of the NucBlue reagent to stain the cell nucleus.
3. Wash the cells three times with warm FluoroBrite DMEM or Leibovitz's L-15 supplemented with 2% dialyzed FBS to remove unbound biotin-CTxB.
4. Incubate the cells with 0.1–0.5 nmol/L streptavidin-QD conjugate in base phenol red-free medium supplemented with 2% dialyzed FBS (*see* **Note 8**) for 1–5 min at 37°C and 5% CO_2.
5. Wash the cells three–five times with warm base phenol red-free medium supplemented with 2% dialyzed FBS. Allow 0–30 min for intracellular QD-GM1 complex accumulation.
6. Add a live-cell imaging buffer and place the glass-bottom dish on the microscope stage. OPTIONAL: to isolate the intracellular QD population, add 10–100 μmol/L QSY-21 quencher [22] to the live-cell imaging buffer (*see* **Note 9**).
7. Locate the sample, set the appropriate acquisition parameters, and begin acquisition as described in Steps 6–9 of Sect. 3.1. Example images are shown in Fig. 2.

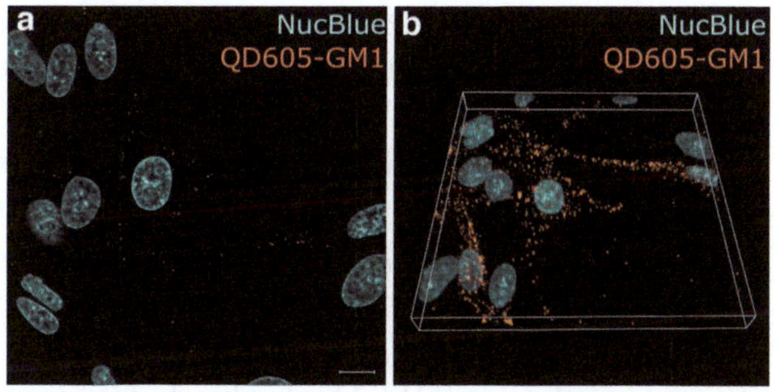

Fig. 2 Example images of QD-streptavidin conjugate labeling of cell surface GM1 molecules. (**a**) A single *z*-plane is showing QD605 distribution with respect to nuclear NucBlue signal (cyan). (**b**) A z-stack rendered in maximum intensity projection (MIP) mode in Imaris shows 3D distribution of QD605 puncta and NucBlue nuclear signal. Scale bar: 10 μm

3.4 Time-Lapse Image Stack Deconvolution and Drift Correction

In this section, directions are provided for iterative image restoration of the acquired *z*-stacks via 3D deconvolution [23] and subsequent drift correction using the Fast4Dreg ImageJ plugin [24]. Deconvolution is a well-established technique to restore images degraded by blur and noise; it is particularly useful in high-resolution volumetric data acquisition where every photon is precious.

1. Launch NIS-Elements AR software.
2. Navigate to the *File* tab in the top menu bar, select *Open File* from the drop-down menu, and open one of the acquired ND2 files to be deconvolved.
3. Navigate to the *Deconvolution* tab and select *3D Deconvolution* from the drop-down menu.
    1. Under the *General* tab, select *Blind* or *Richardson-Lucy* from the Deconvolution Method drop-down menu (*see* **Note 10**). Deconvolution parameters (Modality, Magnification, Pinhole Size, NA, Immersion, Ref. Index, Calibration, *Z*-step) should auto-populate using the metadata contained in the ND2 file (*see* **Note 11**).
4. Choose *Automatic* noise level from the drop-down menu and enter 20 iterations as a starting point. For Richardson-Lucy deconvolution, auto-stopping may be enabled.
5. For the output, select *Create New Document.*
6. Click *Deconvolve* to begin 3D deconvolution. When the processing is complete, save the resulting file in the ND2 format.
7. Launch ImageJ and import the deconvolved ND2 file using Bio-Formats plugin by navigating to *File* > *Import* > *Bio-Formats* (*see* **Note 12**).

8. OPTIONAL: Prepare individual channels for drift correction by first splitting the channels by using *Image > Color > Split Channels.*
9. Save the resulting individual channels as TIFF files (*File > Save As > Tiff...*) and close all open images.
10. Navigate to *Plugins > Fast4Dreg > time estimate + apply* and open time estimate + apply. The user interface opens.
11. In the user interface, add individual TIFF files or a folder containing TIFF files for batch drift correction.
12. Select the *xy-drift correction* option and use the default settings for projection type, time averaging, and maximum expected drift. Select the *previous frame (better for live)* setting for the reference frame.
13. Select the *z-drift correction* option and use the default settings for projection type, time averaging, and maximum expected drift. Select the *previous frame (better for live)* setting for the reference frame.
14. Click OK to start processing.
15. When the script has completed the process, the output subfolder will contain the following files: corrected images, drift plots, drift tables, and a settings file, which can be used to correct the drift of the other channel (*see* **Note 13**).

3.5 Trajectory Reconstruction and Analysis

In this section, directions are given for QD trajectory reconstruction in drift-corrected, deconvolved time-lapse z-stacks using a popular, open-source ImageJ plugin TrackMate [25] and commercial Imaris software.

1. Launch ImageJ 1.54 g and import the drift-corrected, deconvolved TIFF file using Bio-Formats plugin by navigating to *File > Import > Bio-Formats.*
2. Navigate to *Plugins > Tracking* and select *TrackMate* (v 7.14 used here).
3. If the image is properly calibrated, the correct xyz voxel dimensions will be displayed in the initial user interface (*see* **Note 14**). Click *Next.*
4. Select the *LoG detector* from the drop-down list, which will apply the Laplacian of Gaussian filter to the input images in the Fourier space.
5. Enter 0.6 microns for estimated object diameter, five for quality threshold, and select *Sub-pixel localization.*
6. In the open hyperstack (xyzt 4D stack), scroll to the z-plane at the first timepoint that contains QDs appearing as diffraction-limited spots and click *Preview* in the TrackMate user

interface. Once multithreaded detection is complete, the found spots will be displayed. Adjust quality threshold as needed to improve detection.

7. Click *Next* to start 3D detection at all timepoints and again *Next* after detection is done.
8. Use the *Quality* threshold to filter the number of spots before trajectory reconstruction. Click *Next.*
9. Set the desired filters on detected spots and click *Next.*
10. Select the *LAP Tracker* from the drop-down list. This tracker is based on the Linear Assignment Problem framework first implemented by Jaqaman et al. [26].
11. Enter the appropriate settings for the LAP tracker: 0.6 μm max distance for frame-to-frame linking, 0.6–2 μm max distance and 2–10 max frame gap for track segment gap closing with gap closing allowed (*see* **Note 15**). Click *Next* to start the tracking process.
12. Set the desired filters on the reconstructed tracks. A commonly used filter is the number of spots in tracks to exclude very short, noisy trajectories that decrease statistics robustness. Click *Next* to apply the filter.
13. Click on *Tracks* at the bottom of the screen to compile all features of reconstructed tracks, edges, and spots into ImageJ tables. The resulting tables can then be exported as CSV spreadsheets for further analysis.
14. To being track reconstruction in Imaris 10.2, convert the drift-corrected, deconvolved TIFF file to the IMS format using Imaris File Converter 10.2.
15. Open the converted IMS file in Imaris 10.2 and navigate to the *Surpass* 3D view.
16. Click on the *Spots* icon in the *Surpass* tree side panel to create a new *Spots* object and begin the creation wizard.
17. In the first step, check the *Segment only a Region of Interest* box to segment a sub-region of the volume. Select *Different Spots Sizes (Region Growing)* to apply the region growing if QD spots differ in sizes (single QDs and clusters present in the scene), *Track Spots (over Time)* to reconstruct trajectories, and *Object-Object statistics* to activate the shortest distance statistics in the case of a two-channel dataset (Fig. 3). For instance, the shortest distance between the Spot centroid and the nucleus or plasma membrane isosurface (Surfaces object) will be automatically computed, which is useful for monitoring the trafficking process over time (Fig. 4a).
18. Advance to the next step and select the source channel, which will be used to segment the QD Spots. Enter the *Estimated*

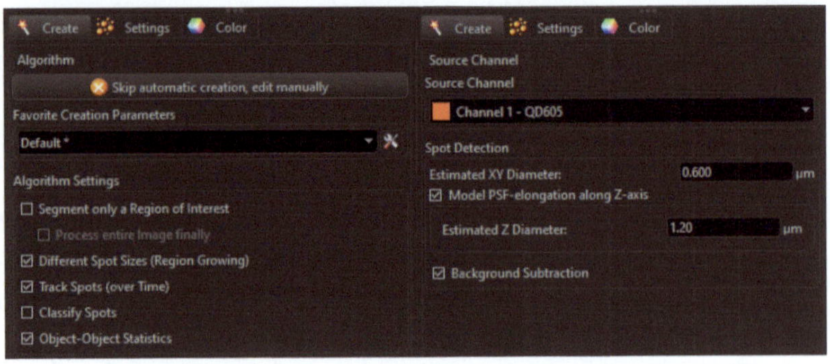

Fig. 3 Imaris Spots creation wizard. The first two steps are shown

XY diameter of 0.6 µm, select *Model PSF-elongation along Z-axis* and enter the *Estimated Z Diameter* of 1.2 µm. Note that If Z > XY diameter, the created shape is a prolate spheroid. Activate *Background Subtraction* (Fig. 3). The intensity of the background-subtracted image is the one of a Gaussian filtered channel (Gaussian filtered by 3/4 sphere radius) minus the intensity of the original channel Gaussian filtered by 8/9 of sphere radius.

19. Next, use the *Quality* filter to adjust the threshold for Spots detection. *Quality* is the intensity value of the detected spot's center in the background-subtracted image.
20. Click on the blue arrow, select *Local Contrast* for Spots growing and *Region Volume* for Spots radius calculation. Adjust the local contrast threshold to increase the size of Spots object for larger QD puncta (i.e., clusters).
21. In the next step, you can manually select spots to delete, duplicate or recenter. This allows you to remove any detected object that you do not want tracked.
22. Next, select the type of *Tracking Algorithm* from the dropdown menu. For most tracking scenarios, *Autoregressive Motion* leads to the accurate reconstruction of Spots movement. Enter the *Max Distance* allowed for frame-to-frame jumps of 1–2 µm and the *Max Gap Size* of 2–10 frames to account for QDs disappearing and reappearing due to blinking.
23. Proceed to the next step to apply filters, such as *Track Duration* or *Track Number of Spots*. Click on the green arrow to complete the Spots Creating and Tracking wizard (Fig. 4b).

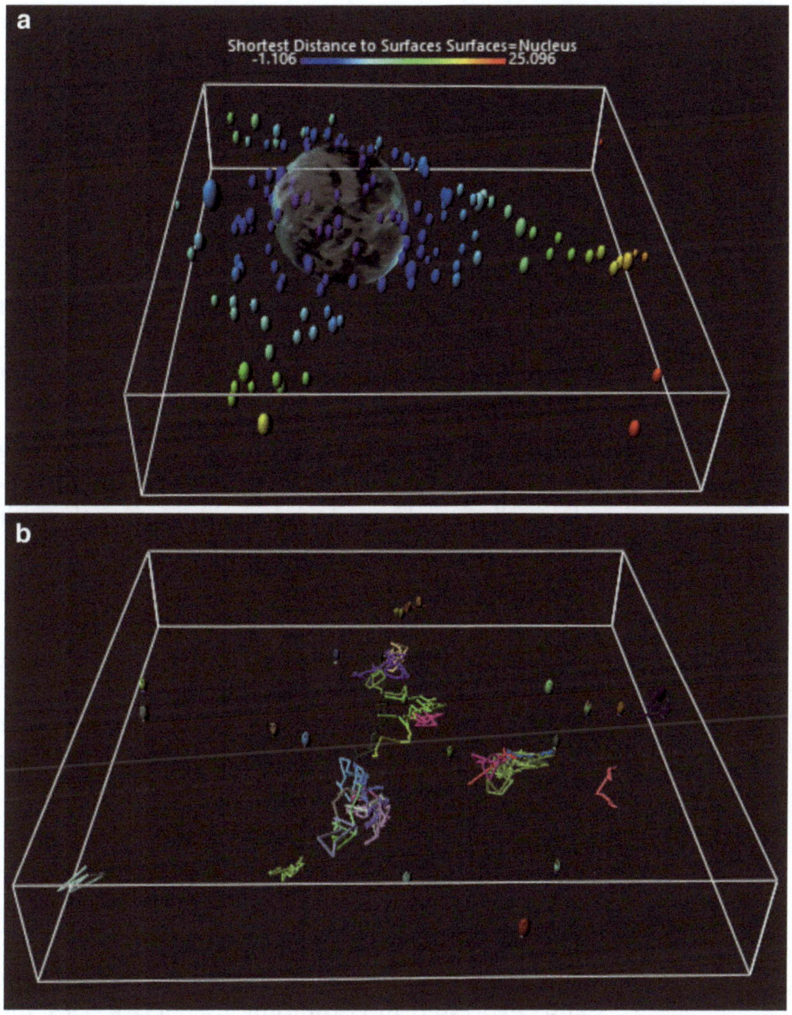

Fig. 4 Examples of three-dimensional reconstruction of QD spot distribution and Qtracker QD trajectories in Imaris. (**a**) The NucBlue signal was reconstructed as a Surfaces object. The QD spots are displayed using color coding based on the automatically calculated proximity of the QD centroid to the edge of the nucleus object. (**b**) Qtracker trajectories were reconstructed from the time-lapse sequence of 50 volumes approximately $30 \times 30 \times 10\,\mu m$

4 Notes

1. Leibovitz's L-15 formulation does not rely on sodium bicarbonate buffering and thus can be used in a CO_2-free environment. Additionally, the L-15 medium uses galactose instead of glucose and may be better tolerated than the standard medium with high glucose concentration (4.5 g/L). Leibovitz's L-15 medium supports the growth of established cell lines

(e.g., Hep-2, LLC-MK2), primary explants of embryonic and adult human tissue, and neuronal cells.

2. Prepare fresh imaging buffers and diluted QD labeling suspensions before each imaging session. Always check buffer pH, as QDs tend to easily aggregate over time, particularly under acidic conditions.
3. It is recommended to start with camera exposure time in the range of 10–50 ms. Longer exposure time will reduce the overall rate of volumetric imaging, whereas shorter exposure time will lead to poor signal-to-noise ratio of individual QDs and can cause the appearance of the spinning disk artifact as arc-shaped streaks.
4. Always add your wash buffer or imaging buffer gently to the side wall of the dish to avoid washing off adsorbed QDs and live cell monolayer. Avoid pipetting growth media and imaging buffers directly onto adsorbed QDs or cells.
5. Keeping the active shutter closed during Z movement is recommended to reduce photobleaching and phototoxicity when imaging live cells. In triggered acquisition in NIS-Elements, some actions during the acquisition are controlled solely by hardware without software interference, resulting in higher frame rates achieved. In *Triggered Piezo Z* mode, a camera triggers the piezo stage.
6. Incubation time and Qtracker QD dose will be cell line dependent. The goal is to avoid accumulation of tens to hundreds of QDs into intracellular vesicles, leading to the appearance of bright large puncta throughout the cytoplasm.
7. GM1 is highly abundant in neuronal tissue and cell lines. It is also expressed in commonly used HEK-293 and HeLa cells model cell lines. However, certain cells, such as Chinese hamster ovary (CHO)-K1, do not synthesize complex gangliosides, including GM1 [27].
8. The concentration of free biotin in commercially available FBS can exceed 50 ng/mL, which is equivalent to approximately 200 nmol/L. Even at 1% FBS in the imaging buffer (2 nmol/L), free biotin can block the available binding sites on QD-streptavidin conjugates. Therefore, it is recommended to use dialyzed FBS with significantly reduced concentration of low molecular weight molecules.
9. QSY-21 absorbs broadly from 550 to 775 nm and serves (concentration range of 1–100 μmol/L) as an efficient quencher of extracellular QDs with emission wavelengths of 605–655 nm [22].
10. According to NIS-Elements documentation, Blind (Iterative Blind) and Richardson-Lucy deconvolution methods reliably

produce top-quality results. Blind deconvolution is more memory demanding as it precisely estimates the shape of the PSF from the parameters specified within the deconvolution dialog window. However, this method may significantly amplify image noise. Richardson-Lucy deconvolution does not iteratively estimate the PSF and may produce lower contrast in the result compared to Blind deconvolution.

11. In case of missing or incorrect metadata in the deconvolution dialog window, right click on the open ND2 image stack and click on the bottom option *Image Properties*. Look for the correct metadata information under the *Image Fields* and *Acquisition Details* tabs.
12. When working with large image stacks in ImageJ, it is recommended to select *Use virtual stack* in the *Bio-Formats Import Options* dialog window to reduce memory usage.
13. Occasionally, Fast4Dreg will not properly correct xyz drift, particularly in the case of sparse QD density in the acquired image series. An alternative is to perform trajectory reconstruction in Imaris and then use a trajectory of an immobile QD in the given dataset to correct the translational and rotational drift of the image. If drift correction fails often, it is then recommended to seed the coverslip of the glass-bottom dish with fluorescent beads to act as fiducial (fixed reference) markers.
14. If the image is uncalibrated in ImageJ, you can modify pixel and time dimensions by navigating to *Image* > *Properties* or using the keyboard shortcut Ctrl+Shift+P.
15. In LAP tracking settings, it is important to test a range for each of the three parameters to avoid false track segment linkages and removal of mobile trajectories with jump sizes larger than the maximum distance of allowed frame-to-frame displacement.

Acknowledgments

Experiments and image analysis were performed in part through the use of the Vanderbilt Cell Imaging Shared Resource (supported by NIH grants CA68485, DK20593, DK58404, DK59637, and EY08126).

References

1. Brandenburg B, Zhuang X (2007) Virus trafficking—learning from single-virus tracking. Nat Rev Microbiol 5(3):197–208. https://doi.org/10.1038/nrmicro1615
2. Liu S-L et al (2020) Single-virus tracking: from imaging methodologies to virological applications. Chem Rev 120(3):1936–1979. https://doi.org/10.1021/acs.chemrev.9b00692
3. Tay A, Melosh N (2019) Nanostructured materials for intracellular cargo delivery. Acc Chem Res 52(9):2462–2471. https://doi.org/10.1021/acs.accounts.9b00272
4. Calebiro D, Nikolaev VO, Persani L, Lohse MJ (2010) Signaling by internalized G-protein-coupled receptors. Trends Pharmacol Sci 31(5):221–228. https://doi.org/10.1016/j.tips.2010.02.002
5. Calebiro D, Koszegi Z (2019) The subcellular dynamics of GPCR signaling. Mol Cell Endocrinol 483:24–30. https://doi.org/10.1016/j.mce.2018.12.020
6. Ha T et al (2022) Next generation single-molecule techniques: imaging, labeling, and manipulation in vitro and in cellulo. Mol Cell 82(2):304–314. https://doi.org/10.1016/j.molcel.2021.12.019
7. Jin D et al (2018) Nanoparticles for super-resolution microscopy and single-molecule tracking. Nat Methods 15(6):415–423. https://doi.org/10.1038/s41592-018-0012-4
8. Lelek M et al (2021) Single-molecule localization microscopy. Nat Rev Methods Primers 1(1):39. https://doi.org/10.1038/s43586-021-00038-x
9. Shen H et al (2017) Single particle tracking: from theory to biophysical applications. Chem Rev 117(11):7331–7376. https://doi.org/10.1021/acs.chemrev.6b00815
10. Breger J, Delehanty JB, Medintz IL (2015) Continuing progress toward controlled intracellular delivery of semiconductor quantum dots. WIREs Nanomed Nanobiotechnol 7(2):131–151. https://doi.org/10.1002/wnan.1281
11. Zhang M-Q et al (2022) Quantum dots tracking endocytosis and transport of proteins displayed by mammalian cells. Anal Chem 94(21):7567–7575. https://doi.org/10.1021/acs.analchem.2c00411
12. Li H et al (2015) Mapping intracellular diffusion distribution using single quantum dot tracking: compartmentalized diffusion defined by endoplasmic reticulum. J Am Chem Soc 137(1):436–444. https://doi.org/10.1021/ja511273c
13. Yang X et al (2022) Probing the intracellular delivery of nanoparticles into hard-to-transfect cells. ACS Nano 16(6):8751–8765. https://doi.org/10.1021/acsnano.1c07648
14. Yang Y-B et al (2020) Single virus tracking with quantum dots packaged into enveloped viruses using CRISPR. Nano Lett 20(2):1417–1427. https://doi.org/10.1021/acs.nanolett.9b05103
15. Zahid MU, Ma L, Lim SJ, Smith AM (2018) Single quantum dot tracking reveals the impact of nanoparticle surface on intracellular state. Nat Commun 9(1):1830. https://doi.org/10.1038/s41467-018-04185-w
16. Rosenthal SJ et al (2011) Biocompatible quantum dots for biological applications. Chem Biol 18(1):10–24. https://doi.org/10.1016/j.chembiol.2010.11.013
17. Lowe AR et al (2010) Selectivity mechanism of the nuclear pore complex characterized by single cargo tracking. Nature 467(7315):600–603. https://doi.org/10.1038/nature09285
18. Zhang Q, Li Y, Tsien Richard W (2009) The dynamic control of kiss-and-run and vesicular reuse probed with single nanoparticles. Science 323(5920):1448–1453. https://doi.org/10.1126/science.1167373
19. Cui B et al (2007) One at a time, live tracking of NGF axonal transport using quantum dots. Proc Natl Acad Sci U S A 104(34):13666–13671. https://doi.org/10.1073/pnas.0706192104
20. Fichter KM, Flajolet M, Greengard P, Vu TQ (2010) Kinetics of G-protein–coupled receptor endosomal trafficking pathways revealed by single quantum dots. Proc Natl Acad Sci U S A 107(43):18658. https://doi.org/10.1073/pnas.1013763107
21. Fujinaga Y et al (2003) Gangliosides that associate with lipid rafts mediate transport of cholera and related toxins from the plasma membrane to endoplasmic reticulm. Mol Biol Cell 14(12):4783–4793. https://doi.org/10.1091/mbc.e03-06-0354
22. Jablonski AE, Kawakami T, Ting AY, Payne CK (2010) Pyrenebutyrate leads to cellular binding, not intracellular delivery, of polyarginine quantum dots. J Phys Chem Lett 1(9):1312–1315. https://doi.org/10.1021/jz100248c
23. Sarder P, Nehorai A (2006) Deconvolution methods for 3-D fluorescence microscopy images. IEEE Signal Process Mag 23(3):32–45. https://doi.org/10.1109/MSP.2006.1628876
24. Pylvänäinen JW et al (2023) Fast4DReg—fast registration of 4D microscopy datasets. J Cell Sci 136(4):jcs260728. https://doi.org/10.1242/jcs.260728
25. Tinevez J-Y et al (2017) TrackMate: an open and extensible platform for single-particle

tracking. Methods 115:80–90. https://doi.org/10.1016/j.ymeth.2016.09.016

26. Jaqaman K et al (2008) Robust single-particle tracking in live-cell time-lapse sequences. Nat Methods 5(8):695–702. https://doi.org/10.1038/nmeth.1237

27. Rosales Fritz VM, Daniotti JL, Maccioni HJ (1997) Chinese hamster ovary cells lacking GM1 and GD1a synthesize gangliosides upon transfection with human GM2 synthase. Biochim Biophys Acta 1354(2):153–158. https://doi.org/10.1016/s0167-4781(97)00117-6

Part III

Labeling and Bioimaging

Chapter 9

Supra-Nanoparticle Assemblies of Quantum Dots for Extracellular Immunolabeling

Ghinwa H. Darwish, Paula Fernández-Gómez, Valle Palomo and W. Russ Algar

Abstract

With their exceptional fluorescence properties, colloidal semiconductor quantum dots (QDs) are powerful tools for bioanalysis and imaging. Examples of these properties include broad absorption spectra, narrow emission spectra, and both superior resistance to photobleaching and higher brightness than fluorescent dyes and proteins. Nevertheless, demanding applications, such as the detection of low-abundance biomarkers, single-particle tracking, and point-of-care diagnostics with low-cost portable devices, will benefit from even higher brightness. With their size-dependent optical properties, amplifying the brightness of QDs is not a simple matter of increasing the nanocrystal size. Instead, multiple copies of QDs must be assembled into a larger nanoparticle (NP) entity. In this chapter, we describe methods for the preparation and characterization of supra-nanoparticle assemblies of QDs (supra-QDs). The supra-QDs feature many individual QDs bound to a silica NP scaffold and are functionalized with dextran to improve colloidal stability. Both pure colors and composite colors of supra-QD are possible. We also describe two methods for the selective immunofluorescent labeling of an extracellular biomarker with supra-QDs: the use of tetrameric antibody complexes (TAC) and covalently linked supra-QD-antibody conjugates.

Keywords Quantum dot, Silica nanoparticle, Dextran, Immunolabeling, Tetrameric antibody complex, Cell imaging

1 Introduction

Colloidal semiconductor quantum dots (QDs) are prominent fluorescent materials in bioanalysis and imaging. Their potential advantages over traditional fluorescent dyes and proteins include superior brightness and resistance to photobleaching, broad absorption spectra paired with narrow and symmetric emission spectra, tunable emission colors, and surface area for chemical and biomolecular functionalization [1, 2]. Even with the exceptional brightness of QDs, some applications require even higher brightness for adequate sensitivity. Examples of such applications include the detection of low-abundance biomarkers and the use of technology-limited portable devices (or

Goreti Pereira et al. (eds.), *Quantum Dots*, Methods in Molecular Biology, vol. 3025,
https://doi.org/10.1007/978-1-0716-5210-7_9,

visual detection) for point-of-care diagnostics. Unlike some varieties of luminescent nanoparticles (NPs), which can be grown larger to increase brightness, the fluorescence properties of QDs are a direct outcome of their small sizes (<10 nm core diameter) and cannot be enhanced by growing larger nanocrystals. Alternative approaches are thus needed to amplify QD properties for improved performance in demanding applications.

Assembling multiple copies of QDs into a larger NP structure is one approach for achieving higher levels of per-particle brightness. Multi-QD assemblies are prepared using a few general strategies: embedding many QDs into carrier microparticles [3–5], assembling many QDs around a central scaffold NP (i.e., supra-QD assemblies)[6–8], and assembling many QDs into a larger NP entity (i.e., super-QD assemblies) [9–12]. For supra-QD assemblies, useful scaffold materials have included silica NPs (SiO_2 NPs), dendrimers, polymer NPs, magnetic NPs, and metal NPs. Among these materials, SiO_2 NPs are particularly favored due to their aqueous compatibility, robustness, transparency, well-established synthetic methods, and numerous possibilities for surface functionalization.

We have prepared supra-QD assemblies based on SiO_2 NP scaffolds to enable immunofluorescent single-cell detection and flow cytometry on smartphone-based imaging platforms [13–15]. We have also shown that the supra-QDs outperform individual QDs in the same applications with research-grade fluorescent microscopes [13, 14]. Our supra-QDs are assembled entirely through affinity interactions: the SiO_2 NPs are functionalized with triethoxy-3-(2-imidazolin-1-yl)propylsilane (IPS), which spontaneously binds to QDs through a combination of electrostatic and coordinate interactions with its imidazoline groups. This facile approach provides supra-QDs that are non-blinking with per-particle photoluminescence (PL) brightness that is enhanced by an order of magnitude [13]. A palette of more than a dozen colors can also be prepared by co-assembling different ratios of red-, green-, and blue-fluorescent QDs to a common SiO_2 NP scaffold, yielding access to the full red–green–blue (RGB) color space, including pinks and purples that are difficult to access with a single size of QD [15]. To improve colloidal stability and reduce nonspecific binding, the supra-QDs are further functionalized with dextran (Dex). Modification of the dextran with pendant 1-(3-aminopropyl)imidazole (API) groups yields spontaneous binding to the assembled QDs. The dextran functionalization also allows tetrameric antibody complexes (TACs) to be used for immunofluorescent labeling with the supra-QDs. TACs consist of an anti-biomarker antibody, an anti-dextran antibody, and two bridging antibody fragments [16–18]. This affinity-driven approach avoids many of the common challenges associated with the immunoconjugation of NPs via covalent chemistries [6, 9, 10, 13, 14, 19, 20]. Nevertheless, the use of carbodiimide coupling (a traditional covalent chemistry) is possible with API-modified carboxymethyl-dextran (CM-Dex) [14].

In this chapter, we provide detailed methodology for the preparation and characterization of supra-QD assemblies, including their functionalization with API-modified dextran and carboxymethyl dextran. Methods for the preparation of pure colors and composite colors based on red-, green-, and blue-emitting QDs are included. Protocols are also provided for the immunofluorescent labeling of cells, whether via TACs or covalent immunoconjugation of the supra-QDs using carbodiimide chemistry. Labeling of human epidermal growth factor receptor 2 (HER2) on the surface of SK-BR3 breast cancer cells is presented as a model system for application of the supra-QDs; however, the materials and methods are applicable to the labeling of other extracellular antigens and other cell types. By summarizing our methods, we aim to facilitate the adoption of these advanced fluorescent materials by other researchers, ultimately expanding their utilization and impact in research and downstream applications.

2 Materials

2.1 Reagents

1. Anhydrous ethanol (EtOH), reagent grade, or better.
2. Deionized water, specific resistance $\geq 18\,M\Omega$ cm.
3. 50 mmol/L borate buffer, pH 8.5.
4. 25 mmol/L and 50 mmol/L HEPES buffer, pH 7.4.
5. Phosphate buffered saline (PBS): pH 7.4, 1.06 mmol/L KH_2PO_4, 155 mmol/L NaCl, 2.97 mmol/L Na_2HPO_4, no $CaCl_2$ or $MgCl_2$.
6. 100 mmol/L carbonate buffer, pH 9.5.
7. Dextran from *Leuconostoc mesenteroides* (Dex; MW 9000–11,000 Da) (*see* **Note 1**).
8. Carboxymethyl dextran from *Leuconostoc mesenteroides*, (CM-Dex; MW 9000–11,000 Da).
9. 0.1 mol/L sodium (meta) periodate (*aq*) ($NaIO_4$).
10. 51 mg/mL sodium cyanoborohydride ($NaCNBH_4$) in deionized water (*see* **Note 2**).
11. Ammonia solution (NH_3), 28%–30% (*aq*).
12. QDs coated with glutathione (GSH) ligands and colloidally stable in water. We have previously reported the reagents and detailed procedure for preparing GSH-QDs [21]. We generally use CdSe/ZnS or CdSe/CdS/ZnS QDs, but we have also used InP/ZnS QDs. In this work, we use QDs with peak PL near 635 nm, but other colors are also suitable.
13. Ethanol/hexane (95:5, v/v) solution.

14. 0.5% w/v bovine serum albumin (BSA) in deionized water.
15. 1 mg/mL of 1-ethyl-3-(3- dimethylaminopropyl)-carbodiimide (EDC) in 50 mmol/L HEPES buffer).
16. 1 mg/mL of *N*-hydroxysuccinimide (NHS) freshly dissolved in 50 mmol/L HEPES (*see* **Note 3**).
17. Anti-HER2 antibody (IgG1, clone HRB2/282).
18. EasySep™ "Do-it-Yourself" positive selection kit II containing TAC reagents.
19. Fixed SK-BR3 cells suspended in PBS buffer (*see* **Notes 4 and 5**).
20. Cell labeling buffer: PBS, 2% w/v BSA, and 1 mmol/L ethylenediaminetetraacetic acid (EDTA), pH 7.4.
21. 0.02% v/v Tween 20 in 25 mmol/L HEPES buffer.

2.2 Equipment

1. Conical centrifuge tubes, 15 mL capacity and 50 mL capacity.
2. Low-binding polypropylene microcentrifuge tubes, 1.7 mL capacity.
3. 96-well microplates (clear, flat bottom; we use half-area, non-treated polystyrene plates) (*see* **Note 6**).
4. Dialysis membrane tubing (regenerated cellulose, 3.5–5 kDa MWCO).
5. Carbon-film-coated copper transmission electron microscopy (TEM) grids. We use 300 mesh.
6. Coverslips: 22 mm × 22 mm, 130–170 μm thickness, borosilicate glass.
7. Microscope slides: 75 mm × 25 mm, 1 mm thickness, white soda-lime glass.
8. Vertical rotating sample mixer.
9. Centrifuge: able to achieve 21,000 × *g* and hold 1.7 mL microcentrifuge tubes.
10. Centrifuge: able to achieve 3000 × *g* and hold 15 mL and 50 mL conical centrifuge tubes.
11. Centrifuge: able to achieve 10,000 × *g* with temperature control and hold 15 mL and 50 mL conical centrifuge tubes.
12. Vacuum centrifuge.
13. Block heater with a metal block for holding 1.7 mL microcentrifuge tubes.
14. Freeze-dryer.
15. Plasma cleaner (air plasma). We use a plasma cleaner with 18 W maximum RF power.
16. Instrument(s) for UV-visible absorbance and fluorescence measurements (*see* **Notes 7 and 8**).

17. Fluorescence microscope for cell imaging (*see* **Note 9**). The microscope will ideally be equipped with filters that enable excitation at wavelengths ≤420 nm and imaging of most of the emission bandwidth from the selected color of QD.
18. *Optional:* fluorescence microscope capable of single-particle imaging (*see* **Note 10**).
19. TEM instrument. We use a TEM operating at an 80 kV accelerating voltage.
20. Nanoparticle tracking analysis (NTA) instrument. We use a model equipped with a 488 nm laser and a 500 nm long pass filter for its fluorescence mode.
21. Dynamic light scattering (DLS) instrument that is configured with phase analysis light scattering (PALS) for zeta potential measurements. We use one equipped with a 640 nm laser.

3 Methods

Figure 1 illustrates and summarizes the preparation steps for assembling QDs onto the surface of silica (SiO_2) NPs to form supra-QDs. SiO_2 NPs are synthesized via a modified Stöber method and are subsequently modified with IPS. The imidazoline group drives the assembly of a monolayer of GSH-coated QDs on the surface of SiO_2 NPs through electrostatic interactions with the anionic GSH ligands on the QDs and through coordinate interactions with the inorganic surface of QDs. The supra-QDs are then functionalized with imidazole-modified dextran, which enhances their colloidal stability, imparts greater resistance to nonspecific interactions with biomacromolecules and cells, and enables the use of TACs (not shown in Fig. 1, vide infra).

Carry out all procedures at room temperature unless otherwise specified.

3.1 Synthesis of SiO_2 NPs

Various sizes of SiO_2 NPs are commercially available or, as described below, can be synthesized using modified Stöber method [22–25].

1. Add 66.7 mL of anhydrous ethanol and 3.3 mL of deionized water to a 250 mL Erlenmeyer flask.
2. Add 4.25 mL (0.020 mol) of tetraethyl orthosilicate (TEOS, 98%) to the ethanol and water mixture (*see* **Note 11**).
3. Add 4.20 mL (0.07 mol) of NH_3 (*aq*) (28%–30%).
4. Stir the mixture for 48 h with a stir bar on a magnetic stirrer set at ~700 rpm.
5. Transfer the SiO_2 NPs suspensions into two 50 mL conical centrifuge tubes.

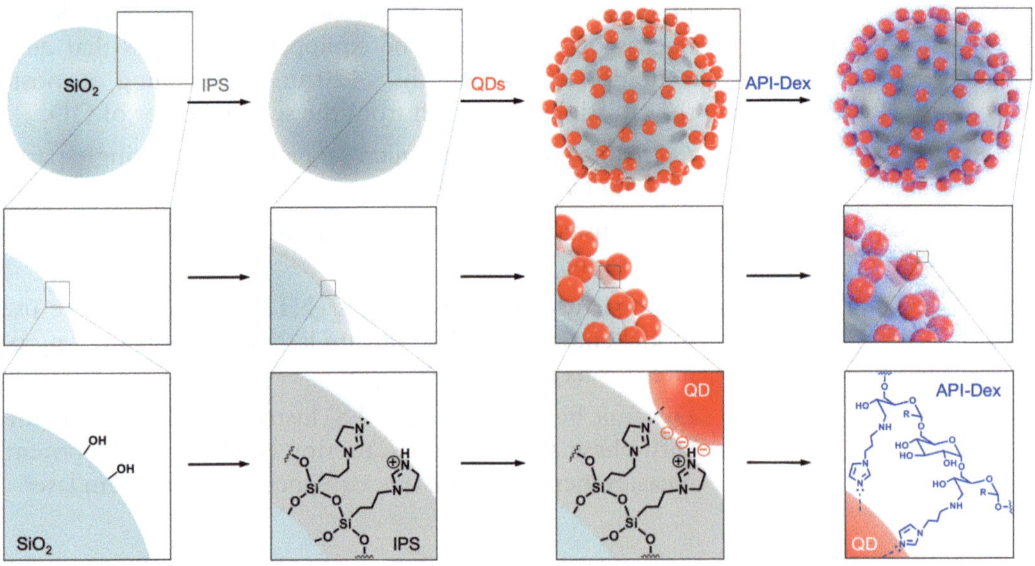

Fig. 1 Schematic representation for the preparation of self-assembled supra-QDs. SiO_2 NPs are modified with IPS, where its imidazoline groups coordinate ZnS-shelled QDs coated with GSH ligands (not shown) and imidazolinium groups bind electrostatically. The QDs are subsequently coated with API-modified dextran, where the imidazole group also coordinates the QDs. The SiO_2 NPs and the QDs are drawn approximately to scale, whereas the IPS and dextran molecules are not to scale. Figure reproduced with permission from [13]. Copyright 2020 American Chemical Society

6. Set the centrifuge temperature to 15 °C.
7. Centrifuge the SiO_2 NPs suspensions at 10,000 ×*g* for 20 min at 15 °C.
8. Carefully draw off the supernatant off the pellet of precipitated SiO_2 NPs and discard. A large translucent white pellet should remain.
9. Redisperse the SiO_2 NPs in 30 mL of EtOH.
10. Repeat Steps 7–9 at least twice (a minimum total of three centrifugation and wash steps).
11. Determine the size and approximate concentration of SiO_2 NPs using NTA (see Sect. 3.7.2).
12. Store at 4 °C. Do not Freeze.

3.2 IPS-Functionalization of SiO_2 NPs

1. Mix 100 μL of the prepared SiO_2 NPs (*ca.* 3 nM; see Sect. 3.1) with 100 μL of triethoxy-3-(2-imidazolin-1-yl)propylsilane (IPS, >97%) (50% v/v) diluted in EtOH (*see* **Note 12**).
2. Add 10 μL of NH_3 (*aq*, 28%–30%).
3. Mix the reaction on a vertical rotating sample mixer for 2.5 h.
4. Centrifuge the solution at 10,000 ×*g* for 5 min.

5. Carefully draw off the supernatant off the pellet of IPS-functionalized SiO_2 NPs and discard. The pellet is typically colorless and translucent and may be difficult to see.
6. Redisperse the IPS-functionalized SiO_2 NPs in 100 µL of borate buffer (50 mM, pH 8.5).
7. Repeat Steps 4–6 at least twice (a minimum total of three centrifugation and wash steps).
8. Measure the zeta potential (see Sect. 3.7.1) to confirm the success of the IPS modification.

3.3 Preparation of Single-Color Supra-QDs

1. Mix 100 µL of IPS-functionalized SiO_2 NPs (*ca.* 3 nM; see Sect. 3.2) with 100 µL of GSH-coated QDs (2–3 µM in borate buffer) (*see* **Note 13**) [21].
2. Let the mixture stand for 1 h in the dark.
3. Centrifuge the solution at 14,000 ×*g* for 5 min.
4. Carefully draw off the supernatant off the pellet of supra-QDs. The supernatant can be discarded or set aside for further characterization (*see* **Note 14**). To avoid disturbing the pellet, check the tube for visible fluorescence under UV light in a dark box. Depending on the color of the QDs, the pellet may also be visible under ambient light.
5. Redisperse the pellet in 100 µL of borate buffer.
6. Repeat Steps 3–5 at least twice (a minimum total of three centrifugation and wash steps).
7. Measure the hydrodynamic size and estimate the concentration of supra-QDs, and confirm the assembly of QDs onto the IPS-functionalized SiO_2 NPs, using the scattering and fluorescence modes of NTA (*see* **Note 15**).

3.4 Preparation of Multicolor Supra-QDs

The preparation of supra-QDs with mixtures of different ratios of red, green, and blue (RGB) colors of QD assembled on single SiO_2 NPs enables access to the full RGB color space, including colors that are difficult to obtain from individual QDs, such as magenta, pink, and white.

1. Mix together volumes of stock solutions of GSH-coated QDs (*ca.* 2 µM stock concentration in borate buffer) prepared with blue (QD450), green (QD540), and red (QD650) emission, to a total volume of 12 µL. The volume of each primary color of QD is optimized empirically to obtain the desired color (*see* **Note 16**).
2. Combine 2.5 µL of IPS-functionalized SiO_2 NPs (*ca.* 150 nM) with the 12 µL of multicolor QD suspension prepared in Step 1.
3. Add 25.5 µL of borate buffer to the mixture.
4. Follow Steps 2–7 from Sect. 3.3.

3.5 Synthesis of Imidazole-Modified Dextran

The synthesis of 1-(3-aminopropyl)imidazole (API)-modified dextran is summarized in Fig. 2.

1. Dissolve 0.5 g of Dex or CM-Dex (~10 kDa, ~50 μmol of polymer chains or ~3.1 mmol of anhydrous glucose monomers) in 20 mL of deionized water in a 40-mL glass vial (*see* **Note 1**).
2. Add 3.1 mL of freshly prepared $NaIO_4$ (*aq*) (0.10 M, 0.31 mmol) to the dextran solution (Fig. 2, Step 1).
3. Cover the glass vial with aluminum foil to protect from light and stir overnight at 4 °C using a stir bar and a magnetic stirrer.
4. Using 3.5 kDa-MWCO tubing, dialyze the sample against 1.0 L of deionized water for approximately two days (*ca.* 45–50 h) with three water changes at regular intervals.

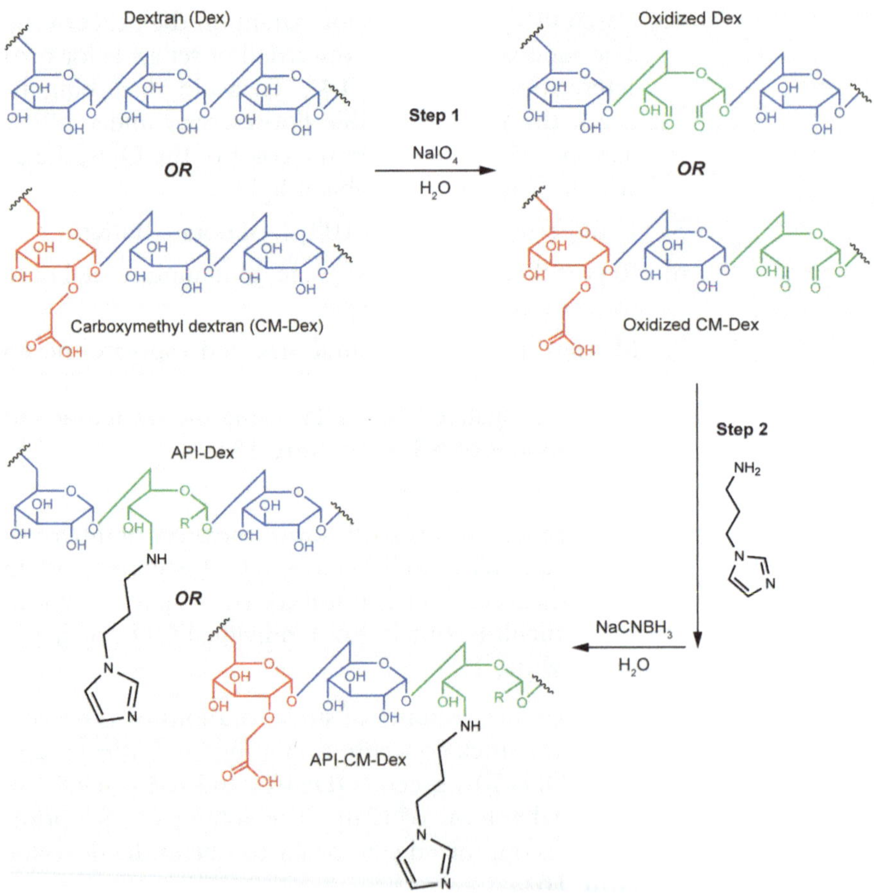

Fig. 2 Synthesis scheme for API-modified dextran or carboxymethyl dextran. The R group may correspond to another secondary amine linkage to API, an unreacted aldehyde, or a cyclized aldehyde with the secondary amine of the shown API modification. The structures are illustrative and not intended to represent the actual stoichiometry or relative positions of the CM and API modifications along the dextran chain

5. Freeze-dry the oxidized dextran to obtain a fluffy white solid.
6. Dissolve 0.10 g of oxidized dextran (~0.62 mmol of anhydrous glucose monomers) in 2.0 mL of deionized water in a 20-mL glass vial.
7. Add 19.7 μL of API (0.165 mmol) to the oxidized dextran solution.
8. Stir the reaction mixture for 2 h using stir bar and a magnetic stirrer.
9. Add 0.25 mL of the prepared $NaCNBH_4$ (*aq*) (51 mg/mL, 0.81 mmol) solution to the API and oxidized dextran solution (*see* **Notes 2 and 17**) (Fig. 2, Step 2).
10. Stir the reaction overnight using a stir bar and a magnetic stirrer.
11. Transfer the full volume of the dextran solution into 50 mL conical centrifuge tube containing 30 mL of ethanol/hexane (95%/5% v/v) solution to induce precipitation (*see* **Note 18**).
12. Centrifuge the precipitated dextran from Step 11 at 3000 ×*g* for 10 min.
13. Carefully draw off the supernatant using a 1 mL micropipette or a Pasteur pipette and discard.
14. Redissolve the pellet in 2.0 mL of deionized water.
15. Repeat Steps 11–14 at least twice (a minimum total of three precipitation and wash steps).
16. Dry the API-modified dextran under reduced pressure in a vacuum centrifuge or using a freeze-dryer.
17. Store the API-modified dextran at 4 °C until needed.

3.6 Dextran Functionalization of Supra-QDs

1. Dissolve 28 mg of API-modified dextran or API-modified CM-Dex in 1.0 mL of carbonate buffer.
2. Mix 40 μL of supra-QD (ca. 3 nM) with 40 μL of the solution from Step 1 in a 1.7 mL microcentrifuge tube.
3. Place the sample in a block heater at 60 °C for 1 h.
4. Wait for the solution to cool to room temperature.
5. Centrifuge the solution at 14,000 ×*g* for 5 min.
6. Gently draw off the supernatant using a 100 μL micropipette. Avoid disturbing the pellet (*see* **Note 19**).
7. Redisperse the pellet of supra-QDs in 100 μL of borate buffer.
8. Repeat Steps 5–7 at least twice (a minimum total of three centrifugation and wash steps). On the final repetition, redisperse the dextran-coated supra-QDs [denoted as supra-(QD-Dex) or CM-dextran-coated supra-QDs (supra-(QD-CM-Dex)] in 40 μL of borate buffer.

3.7 Characterization of Supra-QDs

3.7.1 Zeta Potential

Zeta potential measurements are useful for confirming the successful surface functionalization of SiO_2 NPs with IPS, and for confirming the subsequent assembly of QDs (Fig. 3). For these steps, the surface functional groups change from ionizable silanol groups (anionic, pK_a ~4.5 and ~8.5) [26–28] to ionizable imidazoline groups (cationic, pKa ≥ 10) [29, 30], then to GSH-coated QDs (anionic, the pKa values of the COOH groups are 2.1 and 3.6, and the pKa of the NH_3^+ group is 8.66) [31–33]. The following steps are intended for a Brookhaven Nanobrook Omni instrument but are adaptable to other models of DLS instrument.

1. Prepare separate dilutions of SiO_2 NPs (~1000-fold dilution), SiO_2-IPS (~200-fold dilution), and supra-QD (~200-fold dilution) by diluting the solutions in borate buffer. The total volumes should be at least 2.0 mL. The dilution factors are relative to the as-prepared solutions from earlier steps.
2. Transfer the samples into three separate 4 mL polystyrene cuvettes.
3. Insert the electrodes into the cuvette and measure the zeta potential using the standard operating procedure for the instrument.
4. Perform at least triplicate runs for each sample, ensuring that the count rate is acceptable.

3.7.2 Nanoparticle Tracking Analysis

NTA is useful for estimating the hydrodynamic size distributions and concentrations of SiO_2 NPs and supra-QDs. NTA is also useful for confirming the success of the supra-QD assembly and to estimate the loading density of the QDs per SiO_2 NP as a function of the concentration of GSH-QD solution (*see* **Note 15**). The following steps are intended for a NanoSight NS300 instrument but are adaptable to

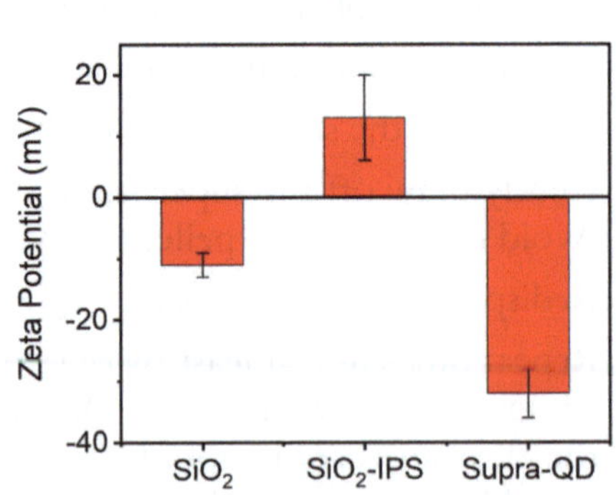

Fig. 3 Example of measured zeta potentials for SiO_2 NPs, IPS-functionalized SiO_2 NPs, and supra-QD assemblies. Data replotted from [13]

other models of NTA instrument (*see* **Note 20**). Figure 4 shows an example of NTA-measured size distribution of the initial SiO_2 NPs and supra-QDs tracked in both scattering and fluorescence modes.

1. Dilute the SiO_2 NP (12,000-fold dilution) or supra-QD (3000-fold dilution) samples in deionized water to a final concentration of 0.2–2 pM. The dilution factors are relative to the as-prepared solutions from earlier steps.
2. Rinse the NTA fluidics with water and ensure that there are no air bubbles.
3. As applicable to the instrument available, set the experiment parameters (e.g., temperature, viscosity, syringe pump flow rate, duration of the recorded video).
4. Load the sample and visualize it in scattering mode. The camera settings will typically need to be adjusted for the supra-QDs to be seen (*see* **Note 21**).
5. Record three videos, each 30 s in duration (*see* **Note 22**).
6. Select a suitable detection threshold to count the particles and exclude any of the background noise and process the data.
7. Switch to fluorescence mode and repeat Steps 4–6 with the supra-QD sample.

3.7.3 Transmission Electron Microscopy

TEM is useful for determining the size distributions of SiO_2 NPs and supra-QDs, for confirming the assembly of QDs onto SiO_2 NPs, and for estimating the number of QDs per SiO_2 NP (Fig. 5) (*see* **Note 23**).

1. Dilute a SiO_2 NP or supra-QD sample with deionized water to a final concentration of 20–40 pM (*see* **Note 24**).
2. Drop-cast 10 μL of the sample on a TEM grid.
3. Let the sample dry overnight in air (*see* **Note 25**).

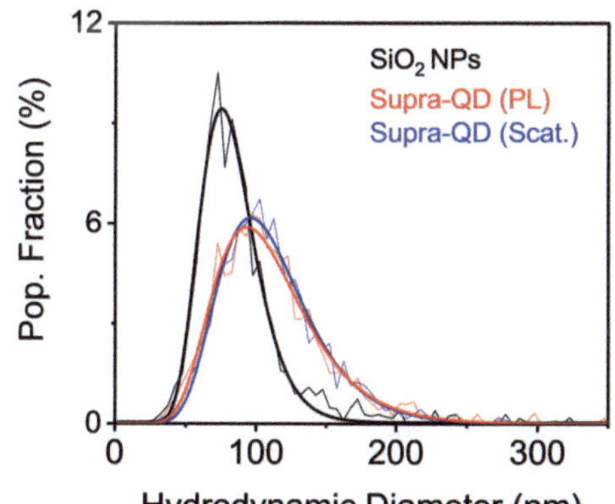

Fig. 4 Examples of NTA-measured size distributions and log–normal curve fits for SiO_2 NPs and supra-QD assemblies, using both scattering (Scat.) and PL modes. Data replotted from [13]

4. Image the grid using TEM at 80 kV (*see* **Note 26**).

3.7.4 Extinction, Excitation, and Photoluminescence Spectra

Extinction spectra (*see* **Note 27**) and PL (*see* **Note 28**) excitation and emission spectra are measured for ensemble-level photophysical characterization of supra-QDs. Typically, the characteristic exciton peaks observed in the extinction spectra of unassembled QDs are obscured when the QDs are assembled on SiO_2 NPs due to the scattering contribution from the silica (Figs. 6A and 7). However, these characteristic peaks remain evident in PL excitation spectra since this measurement is not influenced by scattering. A slight red shift (*ca.* 2–4 nm) in the PL emission spectra is also expected due to energy transfer between smaller and larger QDs that are in close proximity within the assembly (Figs. 6B and 7).

1. Transfer 100 µL of GSH-QD and supra-QD samples into separate wells of a 96-well plate.
2. For extinction spectra: measure from 350 nm to 850 nm.

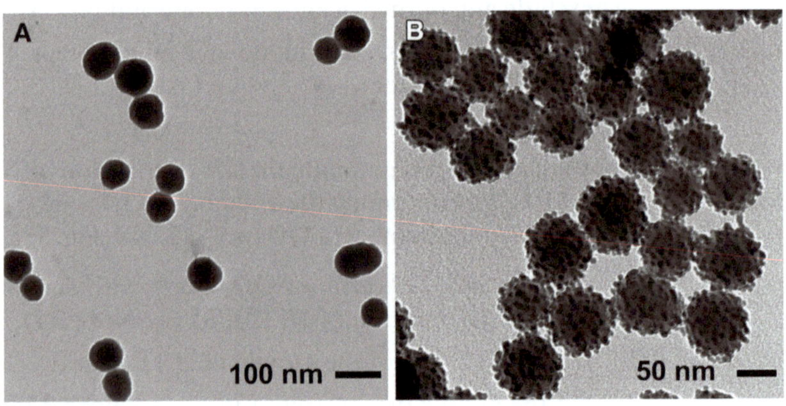

Fig. 5 Examples of TEM images for SiO_2 NPs (**A**) and supra-QD assemblies (**B**)

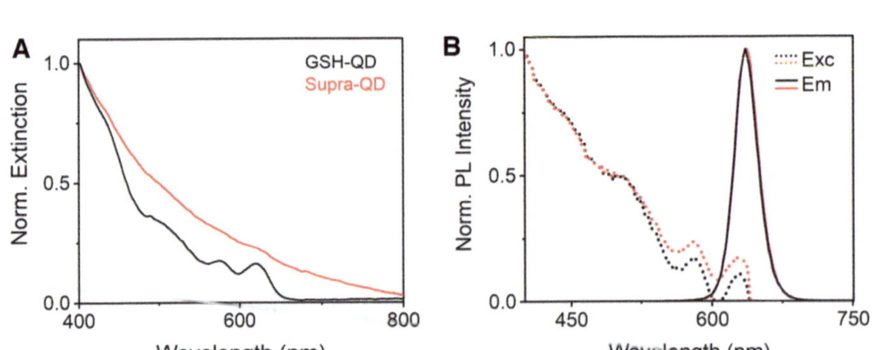

Fig. 6 Examples of ensemble photophysical characterization data. (**A**) Normalized UV–visible extinction spectra and (**B**) normalized PL excitation (Exc.) and emission (Em.) spectra for GSH-QD and supra-QD assemblies. The color legend in panel A also applies to panel B. The data are replotted from [13]

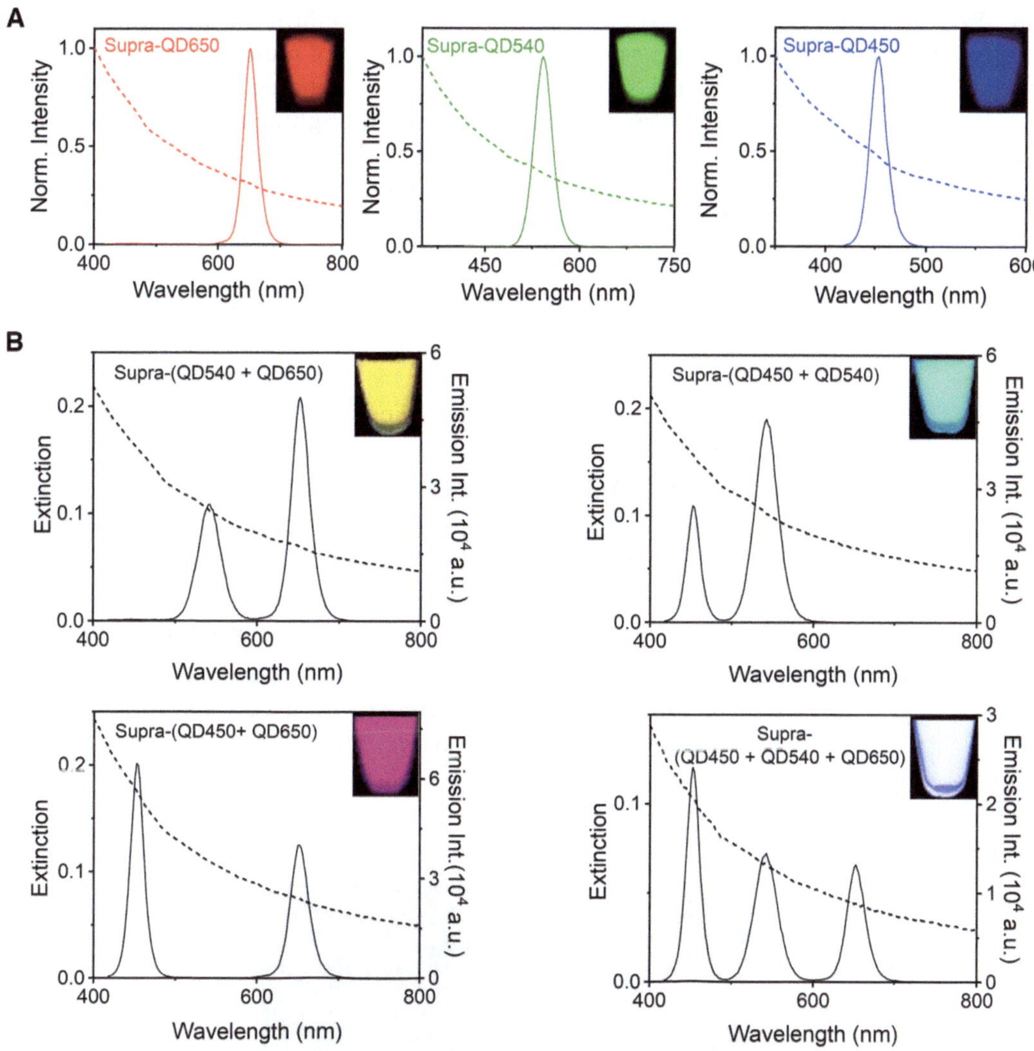

Fig. 7 Examples of ensemble photophysical characterization data for pure and composite colors of supra-QD. UV-visible extinction spectra (dashed line) and PL emission spectra (solid line). (**A**) Primary colors of supra-QDs: supra-QD650 (red), supra-QD540 (green), and supra-QD450 (blue). (**B**) Composite colors of supra-QDs: supra-QD(540 + 650) (yellow), supra-QD(450 + 540) (cyan), supra-QD(450 + 650) (magenta), and supra-QD(450 + 540 + 650) (white). The inset of each graph is a photograph of a microcentrifuge tube containing the indicated supra-QDs, illuminated with UV light in a dark box. Data replotted from [15]. Refer to [15] for the relative ratios of the primary colors of QD needed to yield these composite colors

3. For excitation spectra: set the emission wavelength to the maximum PL emission of the QD and start the scan by varying the excitation wavelength (*see* **Note 29**).
4. For the emission spectra: set the excitation wavelength to 405 nm and start the scan across the desired emission wavelength range (*see* **Note 29**).

3.8 Single-Particle Imaging of Supra-QDs

Single-particle imaging is useful to confirm the formation of supra-QDs, and especially the co-assembly of multiple colors of QDs on individual SiO_2 NPs (Figs. 8 and 9). Single-particle imaging is also useful for comparing the per-particle brightness of supra-QDs relative to GSH-QDs (Fig. 8). However, these measurements are optional for supra-QD characterization, as other complementary techniques, such as NTA and TEM, are often sufficient to confirm the assembly of QDs on SiO_2 NPs.

1. *Optional*: clean glass coverslips (22 mm × 22 mm) using RCA washing [34, 35] or a similar method (e.g., piranha solution) [36].
2. Clean RCA-washed coverslips by using air plasma at 18 W RF for 10 min.
3. Add 100 μL of 0.5% w/v BSA in deionized water onto the coverslip. Use the pipette tip to gently spread the solution, ensuring the entire surface is evenly coated.
4. Place the coverslips in a closed container (e.g., plastic box or a Petri dish) and let stand at room temperature for 30 min.
5. Using a tweezer, carefully hold the coverslip at one edge. With a 1 mL micropipette, remove the excess solution by gently aspirating it from the surface. Then, using 1 mL plastic Pasteur

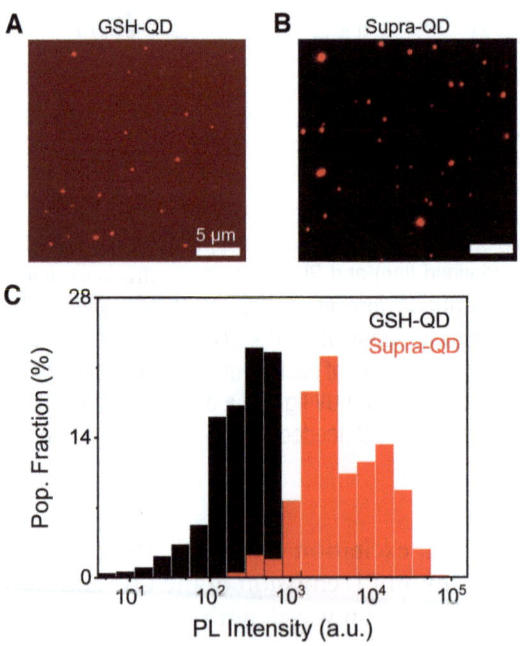

Fig. 8 Example of single-particle photophysical characterization data for supra-QDs. PL images of (**A**) single GSH-QD and (**B**) single supra-QD assemblies. (**C**) Single-particle intensity histograms for GSH-QD and supra-QD assemblies. Scale bars are 5 μm.Histogram data replotted from [13]

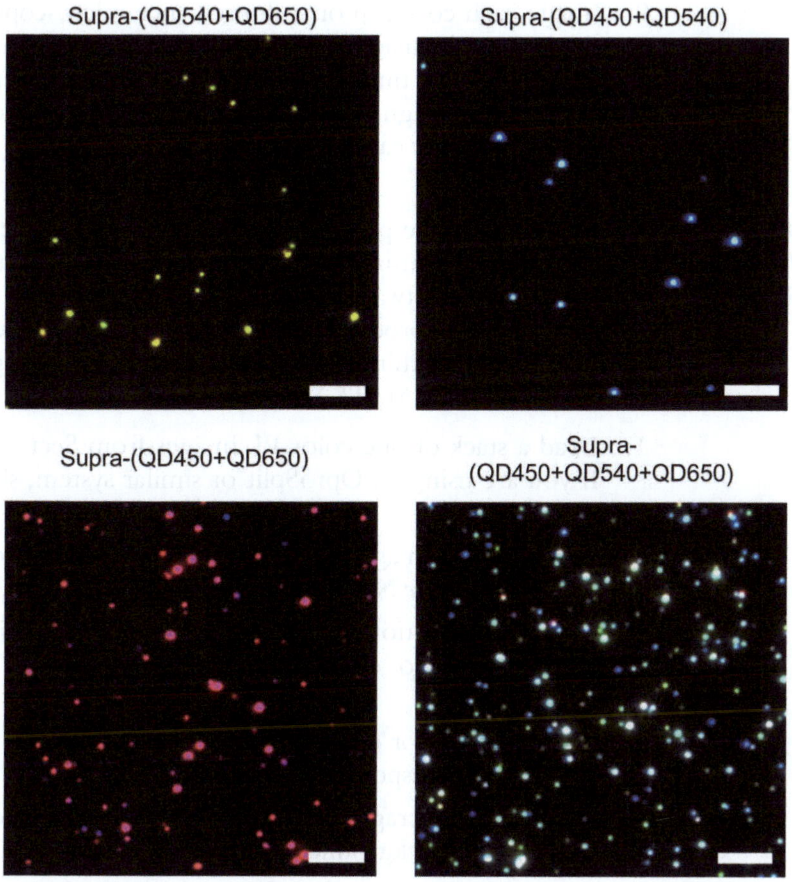

Fig. 9 Examples of single-particle photophysical characterization data for composite colors of supra-QD. The color in the PL images of single particles was calculated from the measured PL intensities in blue, green, and red emission filter channels. Scale bars are 5 μm.Figure adapted with permission from [15]. Copyright 2023 American Chemical Society

pipette, rinse the coverslips ten times with deionized water, ensuring thorough washing. After each rinse, gently tilt the coverslip to help remove excess solution.

6. Add 100 μL of 50–100 pM supra-QDs or 300 pM–1 nM GSH-QD in deionized water or borate buffer (10 μM) to separate BSA-coated coverslips. Use the same technique as described in Step 3 to evenly spread the solution across the surface of each coverslip. Place the coverslips in a closed container at room temperature and let them sit for 1 h.
7. Decant the excess solution and rinse the coverslips ten times with deionized water using a 1 mL plastic Pasteur pipette using the same technique as described in Step 5.
8. Add 100 μL of deionized water to a coverslip prior to imaging.

9. Image each coverslip on a fluorescence microscope capable of single-particle imaging (*see* **Note 10**). Adjust the imaging settings (exposure time, camera gain, and laser power) to ensure having good signal without saturating the camera. Collect image stacks for each sample (*see* **Note 30**).

3.8.1 Analysis of One-Color Single-Particle Brightness for Supra-QDs

The following single-particle image analysis procedure is designed for manual analysis using Fiji (previously known as ImageJ). Other image processing software can be used to perform similar analysis. For more efficient data processing, these analysis steps can be automated by coding scripts within Fiji or using other programming languages, such as Python and MATLAB.

1. Load a stack of one-color PL images from Sect. 3.8 into Fiji. If you are using an OptoSplit or similar system, skip ahead to Sect. 3.8.2
2. Z-project the image frames by average intensity to produce a single image (*see* **Note 31**).
3. Using the selection tool, draw a rectangle or a circle around each particle to define the region of interest (ROI) (*see* **Note 32**).
4. Draw a square or circle of the same dimension as the ROI to define the corresponding background.
5. Measure the average pixel intensities for each particle ROI and the adjacent background using the *measure* function.
6. Subtract the PL intensity of the adjacent background from the PL average of the particle. The subtraction step, all the calculations, and any other data processing steps are done outside of Fiji. Figure 8C shows an example single-particle intensity histogram for QDs and supra-QDs.
7. Exclude any probable aggregates from the analysis (*see* **Note 33**).

3.8.2 Analysis of Multicolor Single-Particle Brightness for Supra-QDs

The OptoSplit III (*see* **Note 34**) can be configured for simultaneous multiwavelength imaging by using three different emission band filters and two dichroic mirrors. This setup allows the camera image to be split into three different wavelength channels, enabling the acquisition of three colors of PL emission simultaneously. In this case, each saved image from the camera contains three wavelength-range-specific copies of the field of view side-by-side-by-side.

1. Load the acquired image stack into Fiji. Each image in the stack from the OptoSplit is a single camera image with three channel-specific images of the same field of view side-by-side-by-side.

2. Z-project the image frames by average intensity (*see* **Note 31**).
3. Draw an ROI around one of the image channels and note the size of the ROI. Create a channel-specific image using the *Duplicate* function (found in the *Image* menu).
4. Repeat the process in Step 3 for the remaining image channels.
5. Align the images from the three wavelength channels using the *Transform* (*affine*) function in the TrakEM2 plugin on Fiji (found in *File* → *New*).
6. Create the composite image from the three channel-specific images using the *Merge Channels* function (found in the *Image* menu → *Color*).
7. To analyze the single-particle brightness, Z-project the image frames by average intensity to produce a single image (*see* **Note 31**). Figure 9 shows example images of single particles of composite colors of supra-QDs.
8. Follow Steps 3–7 in Sect. 3.8.1.
9. Translate the defined particle and background ROIs to the other imaging channels using translate function in the ROI manager.
10. Repeat Steps 3–7 in Sect. 3.8.1.

3.8.3 Analysis of Intensity-versus-Time Trajectories

1. Load a single-color PL image stack into Fiji. Draw a square or circle around each particle to define the ROI.
2. Adjacent to imaged particle, and without overlapping with any particles, draw a square or circle of the same dimension as the ROI in Step 1 to define the corresponding background.
3. Measure PL intensity versus time using the *Time series Analyzer V3* plugin (*see* **Note 35**).
4. Subtract the PL intensity of the adjacent background from the PL average of each particle in each frame in excel sheet. Figure 10 shows examples of intensity-versus-time trajectories for a single GSH-QD (Fig. 10A) and for a single supra-QD assembly (Fig. 10B).
5. Exclude any probable aggregates from the analysis (*see* **Note 33**).

3.9 Immuno-conjugation

3.9.1 Preparation of Tetrameric Antibody Complex

A TAC consists of four antibodies: two antigen-targeting antibodies from one animal species (mouse) and two bridging antibody fragments from another animal species that bind the targeting antibodies together. Here, one end of the TAC is an anti-dextran antibody to bind to dextran-coated supra-QDs [supra-(QD-Dex)] and the other

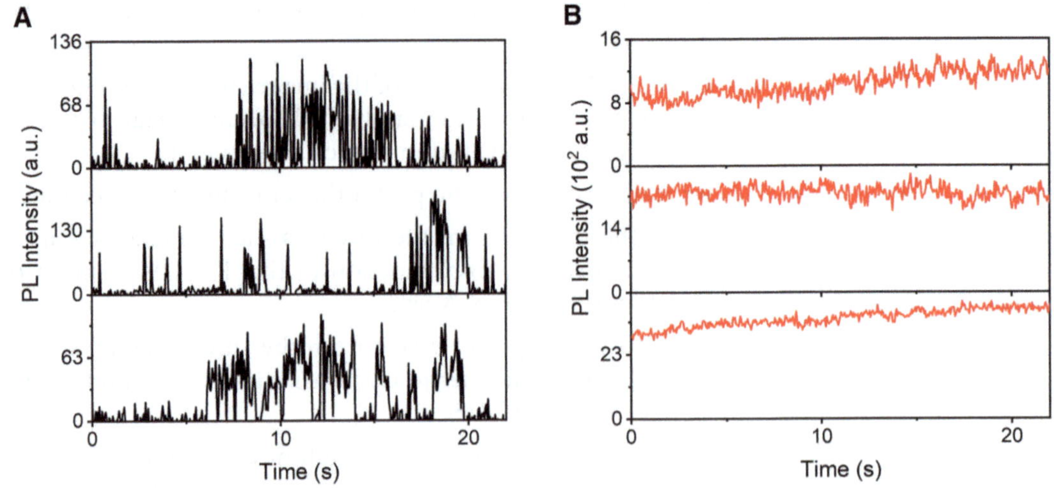

Fig. 10 Representative examples of PL intensity-versus-time trajectories for (**A**) single GSH-QDs and (**B**) single supra-QD assemblies prepared with the same GSH-QDs

end of TAC binds to the specific target cell type (Fig. 11A). An anti-dextran TAC kit is commercially available (*see* **Note 36**) and is used to prepare a 15 μg/mL stock solution of TAC with anti-HER2 in PBS buffer. All the steps follow the manufacturer's protocol for EasySep™ Do-It-Yourself Positive Selection Kit. Anti-HER2 TAC was used for the immunolabeling of the SK-BR3 human breast cancer cell line (*see* **Note 36**).

1. Add 75 μL of 0.2 mg/mL (15 μg) of mouse anti-HER2 into a 1.7 mL microcentrifuge tube.
2. Add 100 μL of kit "component A" to the anti-HER2 solution, followed by adding 100 μL of kit "component B."
3. Incubate the TAC assembly at 37 °C in a block heater (*see* **Note 37**) overnight.
4. Dilute the TAC solution to a final volume of 1 mL with PBS buffer.

3.9.2 Immunoconjugation of Supra-(QD-CM-Dex) with HER-2 Antibody Using EDC/NHS Coupling

1. Dilute 5 μL of ~3 nM of supra-QD-(CM-Dex) in 155 μL of 50 mM HEPES buffer.
2. Add 50 μL of a freshly prepared solution of EDC (~1 mg/mL in 50 mM HEPES buffer).
3. Add 60 μL of a freshly prepared solution of NHS (~1 mg/mL) (*see* **Note 3**).
4. Add 25 μL of 0.02% v/v Tween 20 in 25 mmol/L HEPES buffer.
5. Mix the solution for 30 min in the dark.
6. Centrifuge the solution at 10,000 ×*g* for 5 min.

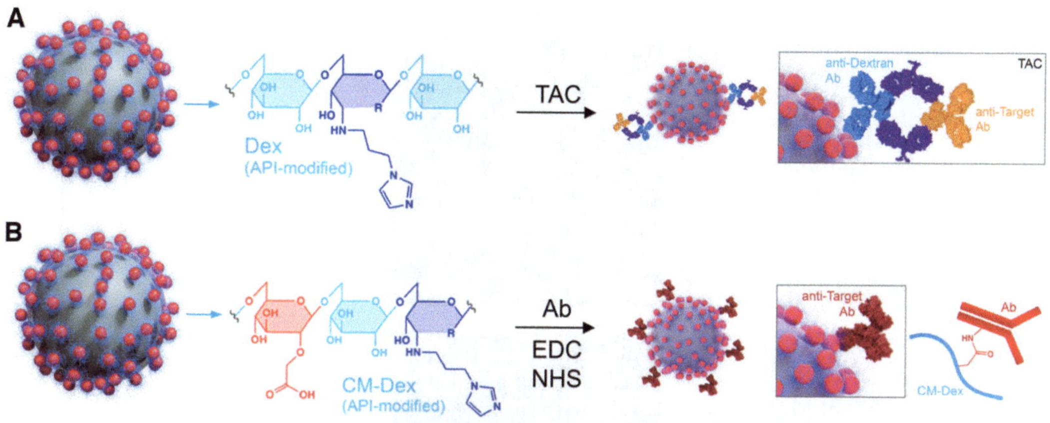

Fig. 11 Strategies for the immunoconjugation of supra-QDs. In this example, the anti-target antibody is anti-HER2, which binds to HER2 on the surface of SK-BR3 cells (not shown). (**A**) Illustration of the immunoconjugation of supra-(QD-Dex) via a TAC of anti-dextran and anti-HER2 antibodies. (**B**) Illustration of immunoconjugation of supra-(QD-CM-Dex) with anti-HER2 antibody via carbodiimide coupling.Figure adapted with permission from [14] which is published under an Open Access License. Copyright 2021 The Authors

7. Carefully draw off the supernatant using a 1 mL micropipette and discard. Redisperse the pellet in 125 μL of 25 mM HEPES buffer.
8. Add 25 μL of 0.02% v/v Tween 20 in 25 mmol/L HEPES buffer.
9. Mix 50 μL of 0.2 mg/mL mouse anti-HER2 with the supra-QD-(CM-Dex)-NHS conjugate and mix in the dark for 1 h.
10. Centrifuge the solution at 10,000 ×*g* for 5 min.
11. Discard the supernatant and redisperse the pellet in 0.5% w/v BSA in 25 mmol/L HEPES buffer.
12. Repeat Steps 10–11 twice (a minimum total of three precipitation and wash steps). On the final repetition, redisperse the supra-QD-(CM-Dex)-(anti-HER2) conjugate in 20 μL 0.05% w/v BSA in HEPES (25 mmol/L) buffer and store at 4 °C for further use. Figure 11B illustrates the immunoconjugation of supra-(QD-CM-Dex) with anti-HER2 antibody via carbodiimide coupling.

3.10 Cell Labeling

Figure 12 shows examples of images and average integrated PL intensities of SK-BR3 cells fluorescently immunolabeled with supra-QDs via anti-HER2-TAC or via carbodiimide-coupled HER2-antibody (and lack of labeling for control samples without antibody).

3.10.1 Cell Labeling with TAC and Supra-QDs

1. Pipette 10 μL of a cell suspension (~2×10^6 cells/mL) into a 1.7-mL microcentrifuge tube.
2. Centrifuge the cells at 100 ×*g* for 5 min.

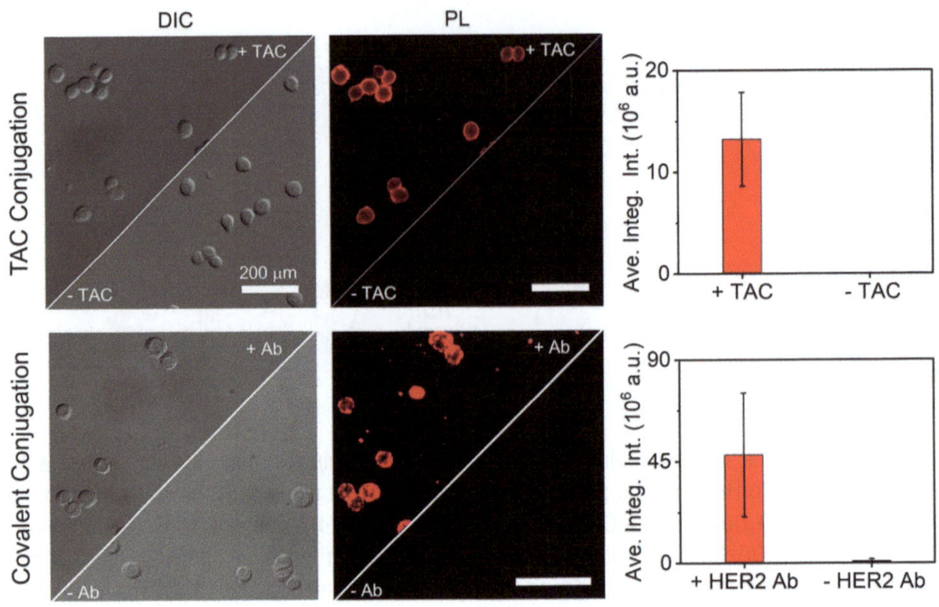

Fig. 12 Examples of fluorescent immunolabeling of HER2-expressing SK-BR3 cells with supra-QDs, using two different immunoconjugation methods: (Top) TAC with supra-(QD-Dex) and (Bottom) HER2-antibody (HER2-Ab) conjugated via carbodiimide coupling with supra-(QD-CM-Dex). The examples include representative differential interference contrast (DIC; left) and PL (Middle) images of the cells, with and without conjugated antibody. All scale bars are 200 μm. The average integrated PL intensity of the cells (N>70) is plotted (Right) for both cases. The error bars are the standard deviations of the average.

3. Avoid disturbing the pellet and carefully remove and discard the supernatant using a 100 μL micropipette, then redisperse the cells in 10 μL of cell labeling buffer.
4. Add 30 μL of TAC-anti-HER2 complex.
5. Mix the solution using a micropipette by gently aspirating and dispensing the solution a few times, then let stand for 30 min in the dark.
6. Dilute the cells with 50 μL of cell labeling buffer.
7. Add 5 μL of ~0.3 nM supra-(QD-Dex). Mix via pipette and let stand for 30 min in the dark. Figure 11A illustrates the immunoconjugation of supra-(QD-Dex) with TAC. The cell being labeled is omitted from the illustration for clarity (*see* **Note 38**).
8. Pellet the cells via centrifugation at 100 ×g for 5 min.
9. Remove the supernatant via pipette and redisperse the cells in 100 μL of cell labeling buffer.
10. Repeat Steps 8 and 9 twice (a minimum total of three centrifugation and wash steps). On the final repetition, redisperse the cells in 10 μL cell labeling buffer.

11. Add 7.5 μL of the sample to a microscope slide, then apply a coverslip.
12. Image on a fluorescence microscope and process the collected images.

3.10.2 Cell Labeling with Supra-QD/HER2 Antibody Conjugates

1. Pipette 25 μL of cell suspension (~2 × 10^6 cells/ mL) into a 1.7 mL microcentrifuge tube.
2. Centrifuge the cells at 100 ×*g* for 5 min.
3. Discard the supernatant and redisperse the cells in 50 μL of cell labeling buffer.
4. Add 20 μL of the supra-QD-(CM-Dex)-anti-HER2 conjugate.
5. Mix via pipette by gently aspirating and dispensing the solution a few times, then let stand for 1 h in dark.
6. Pellet via centrifugation at 100 ×*g* for 5 min.
7. Avoid disturbing the pellet and carefully remove and discard the supernatant using 100 μL micropipette, and then redisperse in 100 μL of cell labeling buffer.
8. Repeat Steps 6 and 7 twice (a total of three centrifugation and wash steps). On the final repetition, redisperse the cells in 20 μL cell labeling buffer.
9. Add 7.5 μL of the sample to a microscope slide, then apply a coverslip (*see* **Note 39**).
10. Image on a fluorescence microscope and process the collected images (*see* **Note 9**).

4 Notes

1. Dextran specified as from *Leuconostoc medenteroides* and *Leuconostoc* spp. has been used successfully. Testing of various dextran molecular weights revealed that lower molecular weights (~6–11 kDa) did not significantly affect QD assembly. However, higher molecular weights retained fewer QDs in the assemblies, perhaps due to an increased number of API groups per dextran chain, which may have outcompeted the surface imidazoline groups for binding with the QDs. Other dextran molecular weights can be used if desired; however, the number of API groups should be adjusted accordingly to minimize competitive interactions.
2. Sodium cyanoborohydride is a potential source of poisonous cyanide and careful handling is, therefore, required. The bottle should only be opened inside a fume hood because hydrogen cyanide gas may form during storage. Always handle

the reagent in a fume hood with proper personal protective equipment. To reduce the risk of inhalation, the solid reagent should be weighed in a sealed container, such as a closed microcentrifuge tube or small glass vial. If a balance is not available in the fume hood, pre-weigh the empty container and weigh the sodium cyanide by difference.

3. It is best to initially dissolve the NHS in 10–20 µL of dimethylsulfoxide (DMSO) before dilution to the final volume with the desired buffer.

4. We cultured SK-BR3 cells in McCoy's 5 A medium, incubated in a 95% air and 5% CO_2 humidified environment at 37°C. The McCoy's culture medium contained phenol red and was supplemented with 10% v/v fetal bovine serum and 1× antibiotic and antimycotic. The cells were cultured in T25 flasks and subcultured every 3–4 days. We anticipate that other standard cell culture procedures will also be suitable and functional.

5. We used formaldehyde for cell fixation. Other cell fixation approaches have not yet been tested and may or may not work. Adhered SK-BR3 cells were detached and suspended by adding trypsin. The suspended cells were then pelleted via centrifugation at 55 ×*g* for 5 min. The supernatant was discarded, and the cells were redispersed in PBS. An equal volume of 4% w/v paraformaldehyde was then added and the cells were mixed by pipette. The cells were incubated with paraformaldehyde for 10 min and then pelleted via centrifugation at 55 ×*g* for 5 min. The fixed cells were then suspended in PBS buffer and kept at room temperature (20–25 °C) until needed.

6. We find that half-area plates with clear bottoms and well walls are often satisfactory for both absorbance and fluorescence measurements. However, in some cases, plates with clear walls will have a problem with fluorescence crosstalk between wells. In such cases, a plate with black well walls should be used.

7. QDs have broad absorption spectra. We frequently use a 405 nm wavelength for excitation to match a laser wavelength that we use for various spectroscopic measurements and for certain applications of the materials. Other wavelengths between the UV and blue regions of the spectrum are also suitable for excitation. A good guideline is that the excitation wavelength is at least 50 nm shorter than the peak emission wavelength for a QD.

8. An instrument that measures UV-visible absorption spectra between 400 and 800 nm with at least 1 nm step size is generally sufficient for characterizing most QDcolors. Similarly, a device capable of fluorescence excitation between 300 and 400 nm, while detecting fluorescence emission between 450 and 800 nm, works well for most QD colors. Although

these instruments are not critical to the procedure, they are valuable for characterizing the QDs and supra-QDs.

9. We use an IX83 inverted microscope featuring a broad-spectrum metal-halide light source, a dichroic mirror turret, and excitation and emission filter wheels, along with a sCMOS camera. Images are captured using 20×/0.75 NA and 60× /1.35 NA objective lenses, and the system is controlled with MetaMorph software. However, none of these components are strictly necessary for the protocol—comparable images can be obtained using microscopes with laser or LED excitation sources, different objective lenses, various camera types (e.g., CCD or EM-CCD), and standard filter cube setups.
10. Images of individual QDs and supra-QDs were captured using an inverted epifluorescence microscope equipped with a 405 nm laser (~21 W/cm^2 at 100% power), a laser-matched dichroic mirror, a 100× oil immersion objective lens (1.45 NA), an image splitter loaded with bandpass filters suitable for the PL to be measured, and an EM-CCD camera. The image splitter is only necessary for imaging supra-QDs composed of multiple colors of individual QDs.
11. The described conditions are to synthesize SiO_2NPs with an average hydrodynamic size of *ca.* 100 nm. The amounts of TEOS and ammonia can be adjusted, as well as the ethanol-to-water ratio, to modify the size of SiO_2NPs [23, 37–40].
12. The amount of IPS required depends on factors, such as the age of the IPS stock, its exposure to air and humidity, and the size and concentration of the SiO_2NPs. However, we recommend that the ethanol-to-water ratio remain constant.
13. In some cases, adding an insufficient quantity of GSH-QDs to IPS-modified SiO_2NPs may result in turbidity and aggregate formation. This issue can be resolved by gradually adding more GSH-QDs until the solution becomes clear.
14. The concentration of QDs can be estimated by measuring their extinction spectra using UV-vis spectrophotometer and applying the Beer–Lambert law to the absorbance at the first exciton peak [41, 42]. After pelleting the supra-QDs via centrifugation, the supernatant can be retained for further analysis. The extinction and emission spectra of the supernatant are useful for estimating the concentration of unassembled QDs. The average number of QDs assembled per SiO_2 NP can then be estimated from the difference in concentration between the original QD suspension and the supernatant, accounting for any dilutions.
15. Since SiO_2 NPs are nonfluorescent and most individual QDs are not sufficiently bright to be reliably detected and sized

by NTA instruments, the successful formation of supra-QDs can be confirmed by detecting particles with the fluorescence mode of NTA. The detected particles should have hydrodynamic diameters and concentrations similar to those observed in scattering mode. A lower concentration suggestions incomplete loading of the SiO_2 NPs with QDs.

16. For composite-color supra-QDs, the total volumes and concentrations of QDs were optimized to saturate the surface of the SiO_2 NPs while leaving a negligible quantity of unassembled QDs remaining in the supernatant. This step is important for obtaining assemblies with colors that are as close as possible to the starting color of the mixed QD solution. However, it is important to note that the color of the assembled QDs may differ slightly from the original mixed QD suspension due to energy transfer between smaller and larger QDs within the supra-QD assemblies.
17. Waste containing sodium cyanoborohydride must be managed with caution and in compliance with institutional guidelines and local regulations due to the toxicity of cyanide solutions and the potential release of hydrogen cyanide gas. All handling and chemical treatment of the waste should be performed in a fume hood while wearing appropriate personal protective equipment. Cyanide in waste solutions can be converted to the less-toxic cyanate using an alkaline bleach (sodium hypochlorite) solution.
18. When precipitating the dextran with a 95% ethanol/5% hexane mixture, we have found that adding the reaction mixture to the ethanol/hexane solution results in more efficient precipitation than the other way around. Adding a few drops of 1.0 M HCl may also help with precipitating API-functionalized dextran.
19. Centrifuging the solution while it is still warm may result in an inefficient pelleting. Allow the sample to cool down to room temperature before centrifugation. A small amount of fluorescence in the supernatant is expected. However, if significant fluorescence is observed, it may indicate that too much API-Dex was used during the dextran coating of supra-QDs, such that API outcompeted the surface imidazoline groups for binding with the QDs. The amount of API-Dex can be adjusted to minimize these competitive interactions.
20. We use a Nanosight NS300 for our NTA measurements. Other NTA instruments, such as ZetaView (Particle Matrix), offer additional characterization capabilities, such as zeta potential measurements. As with zeta potential measurements with a DLS instrument, these data are useful for confirming the successful functionalization of SiO_2 NPs with IPS and the subsequent assembly of QDs on the surface of silica. Some ZetaView models are also capable of fluorescence colocalization measurements, which is useful for composite-color supra-QDs.

21. Camera settings should be optimized to enable tracking of both dimmer and brighter particles. This optimization will ideally minimize false tracking of background noise while avoiding pixel saturation for the brightest particles. Examples of relevant settings for the NS300 include the camera level, shutter speed, gain, and histogram adjustments.
22. We perform our NTA measurements at 25 °C with a sample injection rate of 10 μL/min. The manufacturer of our NTA recommends particle concentrations in the range of 10^7–10^9 particles/mL (17 fM–1.7 pM), with 20–100 particles in the field of view. For our supra-QDs, the optimal concentration was approximately 10^8 particles/mL (~ 0.2 pM). Three videos (30 s duration) are recorded. Recording longer videos can improve the accuracy of the concentration measurements by increasing the number of particles tracked over time, resulting in a more representative dataset.
23. QDs are smaller and more electron-dense than SiO_2 NPs. As a consequence, TEM images show QDs as smaller and darker particles assembled around the larger and less dark SiO_2 NPs.
24. In advance of TEM imaging, the concentration of the supra-QDs is estimated using NTA. The sample is then diluted to 20–40 pM to achieve an optimal particle density per field of view (~30 particles) at 100× magnification. A majority of the particles should be well-distributed without forming clusters or drying on top of each other.
25. To prepare TEM samples, the grids are placed on a piece of clean wax film (e.g., Parafilm brand) to prevent any static-induced displacement or flipping of the grid. After drop-casting the sample, the grids are covered with a Petri dish to protect against dust deposition and left to dry overnight.
26. We have primarily imaged supra-QDs with an 80 kV accelerating voltage for TEM, with 120 kV used on occasion. No noticeable changes in the morphology of the supra-QDs was observed when using higher voltage.
27. "Extinction" refers to the combined effect of light absorption and light scattering by a material. Extinction spectra are measured using a standard UV-visible spectrophotometer (or a plate reader with the same capability) without any change in procedure versus a conventional absorption spectrum.
28. The emission from QDs is commonly described as "photoluminescence" (PL) rather than "fluorescence," which is a specific subtype of PL.
29. When we measure excitation and emission spectra, we typically set the monochromators to use a 2 nm step size and a 5 nm bandwidth. Within reason, these parameters can be increased or decreased with minimal effect on the utility of the data.

30. Optimize the imaging setting (e.g., exposure time, camera gain, and laser power) to ensure both dimmer and brighter particles can be detected. This optimization will ideally minimize the exclusion of dimmer particles while avoiding pixel saturation for the brightest particles. If using μManager software, collect image stacks of 500–1000 frames using the multidimensional acquisition option.
31. The Z-Project function is used to derive a single image from an image stack. To use this function in Fiji, open the image stack and select Image → Stacks → Z Project from the menu bar. A dialog box will appear, where the type of projection can be selected. We typically use *average* option for assessing per-particle brightness because, when blinking or other temporal fluctuation in PL intensity occurs, it better reflects what would be seen at the ensemble level. The drawbacks are that it can be more difficult to identify particles that spend a majority of their time in a dark state, and photobleaching during the measurement will diminish the measured value. The ***maximum intensity*** option for Z-projection does not have these limitations but typically overestimates the particle brightness.
32. To define an ROI around each particle and corresponding background in Fiji, first go to analyze → tools → ROI manager. Using the selection tool, such as oval or square, draw the ROI around each particle of interest. Once the ROI is drawn, click Add in the ROI manager to store it or simply press the shortcut T to add it to the list. This process can be repeated for each particle and background regions.
33. To exclude potential aggregates, the spatial PL intensity distribution for each particle's ROI was plotted, and the full-width-at-half-maximum (FWHM) was measured using the FWHM macro plugin in Fiji. A threshold of 370 nm was then used to exclude potential aggregates of QDs and supra-QDs.
34. The OptoSplit III created three adjacent optical images, each corresponding to a distinct emission filter channel. The splitter was equipped with ET470/40 nm, ET520/40 nm, and ET706/95 nm bandpass filters (center/bandwidth) and T495lpxru-UF2 and ZT640rdc-UF2 dichroic mirrors (495 nm and 640 nm cut-offs, respectively) from Chroma. Continuous illumination was applied to the samples, and images were captured at consistent time intervals.
35. The Time Series Analyzer V3 is a plugin for Fiji that allows for the analysis of intensity changes over time in an image stack. To use it, first install the plugin, if needed. Then,

open an image stack in Fiji, and go to Plugins → Time Series Analyzer V3. A dialog box will appear, where Get Average is selected to calculate the average pixel intensity over time.

36. For our proof-of-concept application, we used the SK-BR3 mammalian breast cancer cell line, which overexpresses HER2 antigen. The mouse IgG1 kappa anti-human HER2 antibody (clone HRB2/282) can be substituted with another mouse-derived IgG1 antibody that is relevant to a different cell type or cell-surface marker of interest. Mouse IgG1 antibody is crucial for ensuring compatibility with the other component of TAC Kit.
37. A heat block is typically used for these procedures, though similar results are expected if a water or oil bath is used at the same temperature.
38. We found that the order of addition, whether immunoconjugating the cells with TAC first followed by the addition of supra-QD or immunoconjugating the supra-QD with TAC and then mixing with the cells, does not substantially affect the immunolabeling of cells.
39. To image immunolabeled SK-BR3 cancer cells, 7.5 μL of the cells were sandwiched between a microscope slide and a glass coverslip. If preferred, coverslips can be sealed onto the microscope slide with a clear top-coat nail polish or adhesive chamber spacer to prevent sample evaporation. However, for this protocol, sealing is unnecessary due to the short duration of the imaging process.

Acknowledgments

The authors thank the Natural Sciences and Engineering Research Council of Canada (NSERC), Canadian Institutes for Health Research (CIHR), Canada Foundation for Innovation (DFI), British Columbia Knowledge Development Fund (BCKDF), and the University of British Columbia (UBC) for support of this research. PF-G is grateful for Mitacs Globalink Research Award for research in Canada. VP acknowledges RYC2019-027489-I and PID2021−128340OA-I00, funded by MCIN/AEI/10.13039/501100011033, FSE "invierte en tu futuro" and "European Union NextGenerationEU/PRTR." The authors thank the UBC bioimaging Facility for access to TEM imaging, the UBC Laboratory for Advanced Spectroscopy and Imaging Research (LASIR) for technical support, and the MacLachlan Group for access to a DLS instrument.

References

1. Algar WR, Massey M, Rees K et al (2021) Photoluminescent nanoparticles for chemical and biological analysis and imaging. Chem Rev 121:9243–9358. https://doi.org/10.1021/acs.chemrev.0c01176
2. Wegner KD, Hildebrandt N (2015) Quantum dots: bright and versatile in vitro and in vivo fluorescence imaging biosensors. Chem Soc Rev 44:4792–4834. https://doi.org/10.1039/C4CS00532E
3. Han M, Gao X, Su JZ, Nie S (2001) Quantum-dot-tagged microbeads for multiplexed optical coding of biomolecules. Nat Biotechnol 19:631–635. https://doi.org/10.1038/90228
4. Liu X, Wu W, Cui D, Chen X, Li W (2021) Functional micro-/nanomaterials for multiplexed biodetection. Adv Mater 33:2004734. https://doi.org/10.1002/adma.202004734
5. Fournier-Bidoz S, Jennings TL, Klostranec JM, Fung W, Rhee A, Li D, Chan WCW (2008) Facile and rapid one-step mass preparation of quantum-dot barcodes. Angew Chem Int Ed 47:5577–5581. https://doi.org/10.1002/anie.200800409
6. Tran MV, Susumu KL, Medintz I, Russ Algar W (2019) Supraparticle assemblies of magnetic nanoparticles and quantum dots for selective cell isolation and counting on a smartphone-based imaging platform. Anal Chem 91:11963–11971. https://doi.org/10.1021/acs.analchem.9b02853
7. Jun BH, Hwang DW, Jung HS et al (2012) Ultrasensitive, biocompatible, quantum-dot-embedded silica nanoparticles for bioimaging. Adv Funct Mater 22:1843–1849. https://doi.org/10.1002/adfm.201102930
8. Wu Q, Chen L, Huang L, Wang J, Liu J, Hu C, Han H (2015) Quantum dots decorated gold nanorod as fluorescent-plasmonic dual-modal contrasts agent for cancer imaging. Biosens Bioelectron 74:16–23. https://doi.org/10.1016/j.bios.2015.06.010
9. Rees K, Darwish GH, Algar WR (2023) Dextran-functionalized super-nanoparticle assemblies of quantum dots for enhanced cellular immunolabeling and imaging. ACS Appl Mater Interfaces 15:18672–18684. https://doi.org/10.1021/acsami.3c00861
10. Rees K, Darwish GH, Bernal-Escalante J, O'Connor KM, Cheong IT, Veinot JGC, Algar WR (2024) Dextran-encapsulated nanoparticles and super-nanoparticle assemblies: preparation from quantum dots, fluorescent polymers, and magnetic nanoparticles for application to cellular immunolabeling. ACS Appl Mater Interfaces 16:64554–64567. https://doi.org/10.1021/acsami.4c14719
11. Guan T, Shen Y, Jiang Z, Liu Y, Aori Q, Koidis A, Yao X, Lei H (2022) One-step self-assembling multiple hydrophobic quantum dots and hydrophilic biomolecules into all-inclusive tags for ultrasensitive immunosensors. Sens Actuators B Chem 373:132720. https://doi.org/10.1016/j.snb.2022.132720
12. Vaishnav JK, Mukherjee TK (2019) Surfactant-induced self-assembly of CdTe quantum dots into multicolor luminescent hybrid vesicles. Langmuir 35:6409–6420. https://doi.org/10.1021/acs.langmuir.9b00357
13. Darwish GH, Asselin J, Tran MV, Gupta R, Kim H, Boudreau D, Algar WR (2020) Fully self-assembled silica nanoparticle–semiconductor quantum dot supra-nanoparticles and immunoconjugates for enhanced cellular imaging by microscopy and smartphone camera. ACS Appl Mater Interfaces 12:33530–33540. https://doi.org/10.1021/acsami.0c09553
14. Xiao Z, Darwish GH, Susumu K, Medintz IL, Algar WR (2022) Prototype smartphone-based device for flow cytometry with immunolabeling via supra-nanoparticle assemblies of quantum dots. ACS Meas Sci Au 2:57–66. https://doi.org/10.1021/acsmeasuresciau.1c00033
15. Darwish GH, Baker DV, Algar WR (2023) Supra-quantum dot assemblies to maximize color-based multiplexed fluorescence detection with a smartphone camera. ACS Sens 8:4686–4695. https://doi.org/10.1021/acssensors.3c01741
16. Wognum AW, Thomas TE, Lansdorp PM (1987) Use of tetrameric antibody complexes to stain cells for flow cytometry. Cytometry 8:366–371. https://doi.org/10.1002/cyto.990080404
17. Thomas TE, Sutherland HJ, Lansdorp PM (1989) Specific binding and release of cells from beads using cleavable tetrameric antibody complexes. J Immunol Methods 120:221–231. https://doi.org/10.1016/0022-1759(89)90246-9
18. Lansdorp PM, Aalberse RC, Bos R, Schutter WG, Van Bruggen EFJ (1986) Cyclic tetramolecular complexes of monoclonal antibodies: a new type of cross-linking reagent. Eur J Immunol 16:679–683. https://doi.org/10.1002/eji.1830160615
19. Rees K, Tran MV, Massey M, Kim H, Krause KD, Algar WR (2020) Dextran-functionalized semiconductor quantum dot bioconjugates for bioanalysis and imaging. Bioconjug Chem 31:861–874. https://doi.org/10.1021/acs.bioconjchem.0c00019
20. Lix K, Tran MV, Massey M, Rees K, Sauvé ER, Hudson ZM, Algar WR (2020) Dextran functionalization of semiconducting polymer dots and conjugation with tetrameric antibody complexes for bioanalysis and imaging.

ACS Appl Bio Mater 3:432–440. https://doi.org/10.1021/acsabm.9b00899
21. Wu M, Petryayeva E, Medintz IL, Algar WR (2014) Quantitative measurement of proteolytic rates with quantum dot-peptide substrate conjugates and Förster resonance energy transfer. Quantum Dots Appl. Biol. Springer 1199:215–239. https://doi.org/10.1007/978-1-4939-1280-3_17
22. Stöber W, Fink A, Bohn E (1968) Controlled growth of monodisperse silica spheres in the micron size range. J Colloid Interface Sci 26:62–69. https://doi.org/10.1016/0021-9797(68)90272-5
23. Rahman IA, Padavettan V (2012) Synthesis of silica nanoparticles by sol-gel: size-dependent properties, surface modification, and applications in silica-polymer nanocomposites—a review. J Nanomater 2012, 132424. https://doi.org/10.1155/2012/132424
24. Gholami T, Salavati-Niasari M, Bazarganipour M, Noori E (2013) Synthesis and characterization of spherical silica nanoparticles by modified Stöber process assisted by organic ligand. Superlattices Microstruct 61:33–41. https://doi.org/10.1016/j.spmi.2013.06.004
25. Meier M, Ungerer J, Klinge M, Nirschl H (2018) Synthesis of nanometric silica particles via a modified Stöber synthesis route. Colloids Surf A Physicochem Eng Asp 538:559–564. https://doi.org/10.1016/j.colsurfa.2017.11.047
26. Ong S, Zhao X, Eisenthal KB (1992) Polarization of water molecules at a charged interface: second harmonic studies of the silica/water interface. Chem Phys Lett 191:327–335. https://doi.org/10.1016/0009-2614(92)85309-X
27. Sulpizi M, Gaigeot M-P, Sprik M (2012) The silica–water interface: how the silanols determine the surface acidity and modulate the water properties. J Chem Theory Comput 8:1037–1047. https://doi.org/10.1021/ct2007154
28. Allen LH, Matijevíc E, Meites L (1971) Exchange of Na+ for the silanolic protons of silica. J Inorg Nucl Chem 33:1293–1299. https://doi.org/10.1016/0022-1902(71)80423-2
29. Kim KH, Martin YC (1991) Direct prediction of dissociation constants (pKa's) of clonidine-like imidazolines, 2-substituted imidazoles and 1-methyl-2-substituted-imidazoles from 3D structures using a comparative molecular field analysis (CoMFA) approach. J Med Chem 34:2056–2060. https://doi.org/10.1021/jm00111a020
30. Grimmett MR (1984) 4.07 - Imidazoles and their Benzo Derivatives: (ii) Reactivity. In: Katritzky AR, Rees CW (Eds.) Comprehensive Heterocyclic Chemistry. Pergamon, 373–456. https://doi.org/10.1016/B978-008096519-2.00075-8
31. Pirie NW, Pinhey KG (1929) The titration curve of glutathione. J Biol Chem 84:321–333. https://doi.org/10.1016/S0021-9258(18)77063-4
32. Vašková J, Ko čan L, Vaško L, Perjési P (2023) Glutathione-related enzymes and proteins: a review. Molecules. 28:1447. https://doi.org/10.3390/molecules28031447
33. Li NC, Gawron O, Bascuas G (1954) Stability of zinc complexes with glutathione and oxidized glutathione. J Am Chem Soc 76:225–229. https://doi.org/10.1021/ja01630a058
34. Kern W, Puotinen DA (1970) Cleaning solutions based on hydrogen peroxide for use insilicon semiconductor technology. RCA Rev 31:187–206
35. Tavares AJ, Petryayeva E, Algar WR, Chen L, Krull UJ (2010) Toward a hybridization assay using fluorescence resonance energy transfer and quantum dots immobilized in microfluidic channels. ProcSPIE. https://doi.org/10.1117/12.876746
36. Chandradoss SD, Haagsma AC, Lee YK, Hwang J-H, Nam J-M, Joo C (2014) Surface passivation for single-molecule protein studies. J Vis Exp. https://doi.org/10.3791/50549
37. Bogush GH, Tracy MA, Zukoski CF (1988) Preparation of monodisperse silica particles: control of size and mass fraction. J Non Cryst Solids 104:95–106. https://doi.org/10.1016/0022-3093(88)90187-1
38. Green DL, Lin JS, Lam Y-F, Hu MZ-C, Schaefer DW, Harris MT (2003) Size, volume fraction, and nucleation of Stober silica nanoparticles. J Colloid Interface Sci 266:346–358. https://doi.org/10.1016/S0021-9797(03)00610-6
39. Gao W, Rigout M, Owens H (2016) Facile control of silica nanoparticles using a novel solvent varying method for the fabrication of artificial opal photonic crystals. J Nanoparticle Res 18:387. https://doi.org/10.1007/s11051-016-3691-8
40. Matsoukas T, Gulari E (1988) Dynamics of growth of silica particles from ammonia-catalyzed hydrolysis of tetra-ethyl-orthosilicate. J Colloid Interface Sci 124:252–261. https://doi.org/10.1016/0021-9797(88)90346-3
41. Jasieniak J, Smith L, van Embden J, Mulvaney P, Califano M (2009) Re-examination of the size-dependent absorption properties of CdSe quantum dots. J Phys Chem C 113:19468–19474. https://doi.org/10.1021/jp906827m
42. Yu WW, Qu L, Guo W, Peng X (2003) Experimental determination of the extinction coefficient of CdTe, CdSe, and CdS nanocrystals. Chem Mater 15:2854–2860. https://doi.org/10.1021/cm034081k

Chapter 10

Precise and Stable Labeling of Quantum Dots with Engineered Antibodies

Victor R. Mann, Cassio C.S. Pedroso and Bruce E. Cohen

Abstract

Quantum dots (QDs) are ideal probes for quantitative bioimaging but require controlled surface functionalization to be useful for imaging specific receptors or other cellular targets. Antibodies (Ab) offer high affinities and specificities against a range of targets, but methods for synthesizing QD–Ab conjugates may offer little control over stoichiometry, orient Ab incorrectly, or be reversible or denaturing. Here, we describe a protocol using SpyCatcher proteins, which spontaneously form covalent bonds with paired SpyTag peptides, to stably link Ab to QD surfaces with control over stoichiometry and orientation. Protocols for immunolabeling of CdSe/CdS QDs with miniaturized Ab, their characterization, and their use in live cell imaging are presented.

Keywords Bioconjugation, Microscopy, Receptor, ScFv antibody, Immunotargeting, Protein engineering, Aqueous passivation

1 Introduction

Optical microscopy enables minimally invasive, high-resolution study of biology, and the specific detection of cellular components is essential for understanding molecular signaling in all living systems [1]. Cell-surface receptors are most often detected by antibodies (Ab) conjugated to fluorophores, and recent Ab selection and engineering technologies have generated selective fluorescent probes against a wide range of receptors [2]. Compared with traditional organic fluorophores, semiconductor quantum dots (QDs) have better photostability, larger optical cross-sections, and tunable emission and excitation spectra. QDs have demonstrated uses in demanding bioimaging experiments, including live-cell single-molecule tracking and photoactivation [3–5]. Like other inorganic nanocrystals [6, 7], most high-quality QDs are synthesized as hydrophobic colloids [8] and must be made aqueous and functionalized to have utility as imaging probes or sensors [9–11]. Expanding the reach of QDs for

Goreti Pereira et al. (eds.), *Quantum Dots*, Methods in Molecular Biology, vol. 3025,
https://doi.org/10.1007/978-1-0716-5210-7_10,

imaging applications requires precise and simplified methods for this surface biofunctionalization.

Ab used in microscopy are typically large (150 kDa) and polymorphic immunoglobulin G (IgG) proteins, with two identical antigen binding sites, spaced up to almost 20 nm apart [12]. Numerous bioconjugation reactions have been developed for synthesizing Ab conjugates, although some of these reactions have been shown to be reversible under physiological conditions [13, 14] or to reduce Ab stability or affinity for ligand [15, 16]. Advances in in vitro selection and protein engineering have produced engineered Ab constructs that are much smaller while retaining antigen affinity and selectivity [2]. Single chain variable fragment (scFv) Abs that are *ca.* 25 kDa (or 3 nm diameter) and are much more amenable to expression in *E. coli* [17].

Traditional bioconjugation reactions between Ab sidechains and organic fluorophores have been adapted for the synthesis of QD immunoconjugates [18–20], but difficulties in controlling stoichiometry, Ab orientation, and stability have raised questions about their utility [21]. Bioorthogonal reactions may be useful in expanding the scope of QD bioconjugation reactions [4, 22], although these reaction often require reactant concentration higher than possible for large IgGs and QD [23, 24]. However, the engineered split protein system SpyCatcher/SpyTag [25, 26], in which the components bind to one another with high affinity before spontaneous formation of an isopeptide bond, and has emerged as a general method for irreversible conjugation of proteins to nanoparticles [17, 27–29].

We describe here the synthesis and characterization of Ab–QD conjugates (Fig. 1), in which QDs are encapsulated within amphiphilic polymers [30, 31] that have bioconjugation handles for attachment of peptides, proteins, ligands, or sensors [4, 5, 22, 32, 33]. SpyCatcher proteins are conjugated to those polymers [27–29],

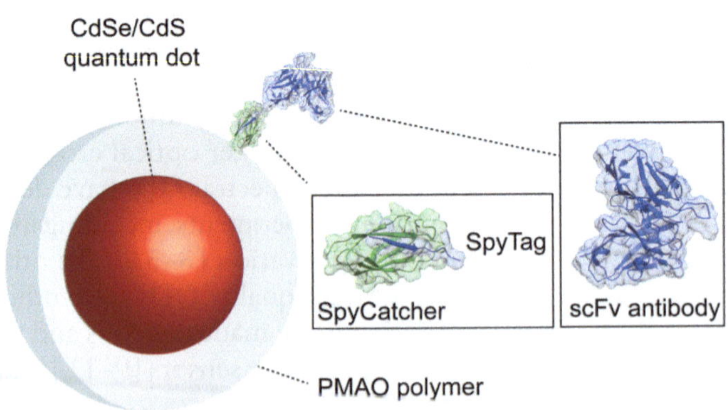

Fig. 1 Schematic of a QD-SpyCatcher-scFv Ab conjugate. CdSe/CdS QD (red), encapsulated in amphiphilic polymers (light gray), conjugated to a single-Cys SpyCatcher (green), and SpyTag-scFv antibody (blue)

enabling simple conjugation of SpyTag-scFv Ab [17]. This offers a modular system in which any hydrophobic nanoparticle can precisely and covalently be labeled with a wide range of bioactive molecules.

2 Materials

2.1 Components for Aqueous Dispersion of Hydrophobic Quantum Dots

All buffers should be prepared with deionized, sterilized water using standard methods. Use anhydrous organic solvents wherever possible.

1. Hydrophobic CdSe/CdS core/shell QDs were synthesized as described [4, 8]. Commercial QDs or other hydrophobic nanocrystals may also be used (*see* **Note 1**).
2. Amphiphilic polymer poly(maleic anhydride-alt-1-octadecene) (PMAO, 10 kDa, Fig. 2) (*see* **Note 2**).
3. 2-(2-(2-methoxyethoxy)ethoxy)ethanamine (m-PEG_3-NH_2).
4. The cross-linker 2-(2-pyridyldisulfide)ethylamine (PDEA, Fig. 2) is prepared as in [17], but other carboxylic acid-to-thiol crosslinkers will work as well.
5. 100 mmol/L sodium borate buffer, pH 8.6.
6. 20 mmol/L HEPES buffer, pH 7.4

2.2 Components for Bioconjugation of Proteins to Quantum Dot Surfaces

1. SpyCatcher S35C is expressed and purified as described in [17].
2. 100 mmol/L HEPES buffer, pH 7.8.
3. 100 mmol/L HEPES buffer, pH 7.5.

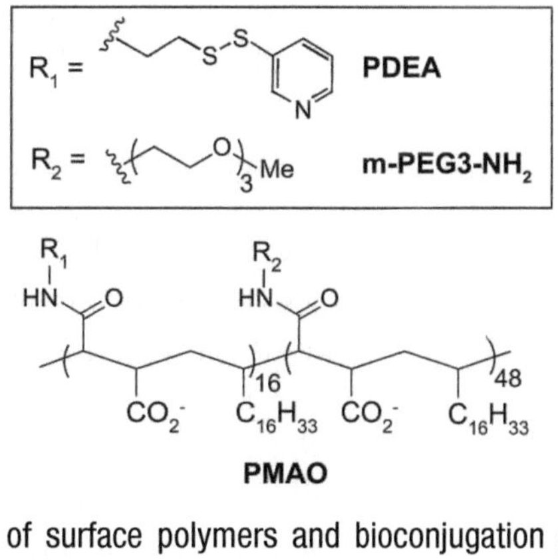

Fig. 2 Structures of surface polymers and bioconjugation groups. Co-maleic anhydride-octadecene polymer (PMAO) modified with PDEA linker and inert PEG amines. Stoichiometries are estimates based on polymer molecular weights, and positions of monomers are random

2.3 Components for Imaging Ab-QD Conjugates with Live Cells

1. Cell culture chamber slides with 8 wells for inverted microscopes (*see* **Note 3**).
2. MDA-MB-231 cells.
3. PBS buffer: 50 mmol/L $NaHPO_3$, pH 7.4, with 100 mM NaCl.
4. 10 mmol/L glucose.

3 Methods

3.1 Preparation of Aqueous PMAO-Quantum Dots

1. Dissolve PMAO polymer (30 mg, 1.33 μmol, 16 monomer units per nm^2 of QD surface area) in 1 mL of acetone and 19 mL of $CHCl_3$. If needed, briefly sonicate in a water bath for full dissolution.
2. Add CdSe/CdS QDs in hexane (250 μL of 8 μmol/L, 2.0 nmol) with gentle stirring, and evaporate off solvents with a gentle stream of N_2 until completely dry (*see* **Note 4**).
3. Resuspend the dry QD/polymer residue in 15 mL of 100 mM sodium borate buffer, pH 8.6 with 1:99 ratio of PDEA (1.3 mg, 5.7 μmol) to m-PEG_3-amine (92 μL, 564 μmol) (*see* **Note 5**).
4. Bath sonicate the suspension for 4 h. Remove excess polymer by five rounds of centrifugal dialysis (100 kDa MWCO, 4000 ×*g*). Purify aqueous QDs of excess polymer by dialysis (10 mL, 100 kDa MWCO) of 3 × 500 mL of 20 mM HEPES buffer, pH 7.4 (*see* **Note 6**).
5. Concentrated the sample by spin dialysis (100 kDa MWCO) to a final of ca. 2 μmol/L in 1 mL (*see* **Note** 7).

3.2 Bioconjugation of Proteins to Quantum Dot Surfaces

1. Express and purify SpyCatcher S35C as described in [17] or other single-cysteine proteins or peptides. Remove any excess of reducing agent by centrifugal dialysis (3 kDa MWCO, 4000 ×*g*) (*see* **Note 8**).
2. Combine PMOA-encapsulated QDs (2 μmol/L, 200 μL) in 100 mmol/L HEPES, pH 7.8 with an aliquot of SpyCatcher S35C (200 μmol/L, 100 μL) in a 1.5 mL centrifuge tube, vortex well, and mix overnight at 4°C on a rotary mixer.
3. Dilute the reaction mixture to 4 mL with 100 mM HEPES, pH 7.5, and remove excess SpyCatcher protein by centrifugal dialysis (100 kDa MWCO, 4000 ×*g*), washing with 4 × 4 mL of buffer.
4. Use the flow-through to determine the degree of bioconjugation using the absorbance of tracer molecule pyridine-2-thione, which is released when a thiol conjugates to PDEA (Fig. 3). The concentration of pyridine-2-thione can be determined

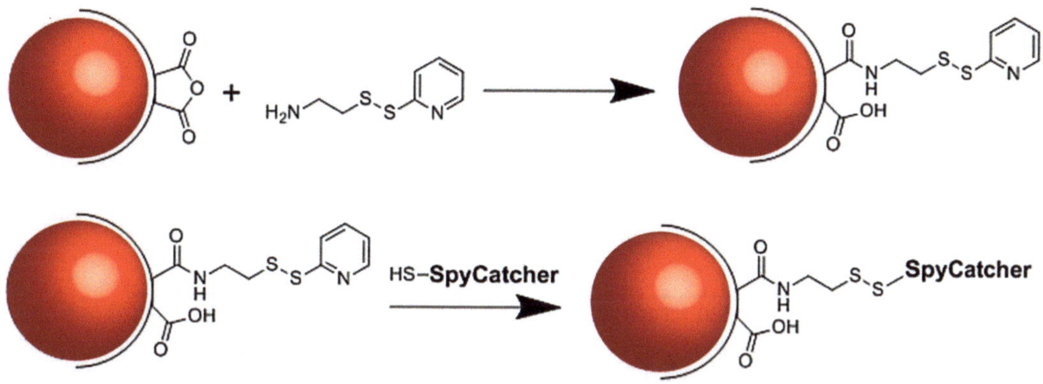

Fig. 3 Reaction of PMAO-coated QDs with PDEA to display activated disulfides (top). Subsequent disulfide exchange (bottom) conjugates single-Cys SpyCatcher proteins to QD surface polymers

from its absorbance at 343 nm, using its extinction coefficient of 8080 $M^{-1}cm^{-1}$.

5. Dilute the retentate to 400 μL in 100 mmol/L HEPES, pH 7.5, and store the QD–protein conjugate at 4°C if not being used immediately.

3.3 Coupling of Engineered Antibodies to Quantum Dot Surfaces

1. Mix SpyCatcher-coated QDs (100 μL, 1 μmol/L, 100 pmol) and SpyTag-scFv antibody (8.9 μL, 22.6 μmol/L, 201 pmol, ~2 per QD) gently overnight. This may be used for microscopy without purification.
2. If purification is desired, remove trace free scFv antibody by centrifugal dialysis (100 kDa MWCO, 4000 ×*g*), washing with 2 × 4 mL of 100 mM HEPES, pH 7.5.
3. The number of scFv Ab per QD may be tuned by varying the volume of added SpyTag-scFv solution (Fig. 4).
4. To verify the presence of Ab on the QD surface, acquire fluorescence spectra with 270 nm excitation prior to and after conjugation, with the difference curve showing a tryptophan emission peak at ~340 nm (*see* **Note 9**). A calibration curve of known concentrations of purified SpyTag-scFv may be used to determine the concentration of Ab conjugated to the nanoparticle.

3.4 Imaging scFv-QD Conjugates with Live Cells

1. Before addition to MDA-MB-231 cells, sterilize scFv-QD conjugates by passage through a 0.22-μm filter.
2. Grow cells (*see* **Note 10**) in chamber slides (*see* **Note 3**) to 80%–90% confluency. Aspirate the wells of media and wash with 2 × 500 μL complete media without phenol red.
3. Dilute QD conjugates to 10 nmol/L in media without phenol red and add to the cells (*see* **Note 10**). After 1 h in a 37°C

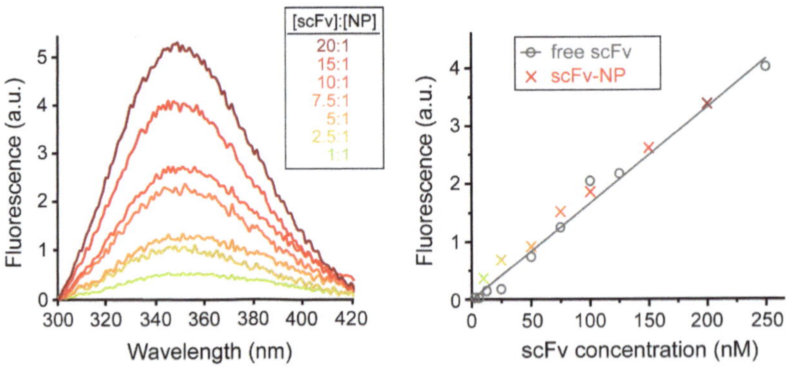

Fig. 4 Quantification of scFv Ab conjugated to SpyCatcher-coated QDs by Trp fluorescence. (Left) Trp emission spectra of 10 nM scFv-NP conjugates. Spectra are corrected for SpyCatcher Trp emission by subtracting 10 nM SpyCatcher-NP emission. (Right) Quantification of scFv per NP (×symbols, 1 to 20 per NP) by comparison with a free SpyTag-scFv standardization line (gray circles and line). All emission spectra were integrated from 300 to 420 nm with 290 nm excitation

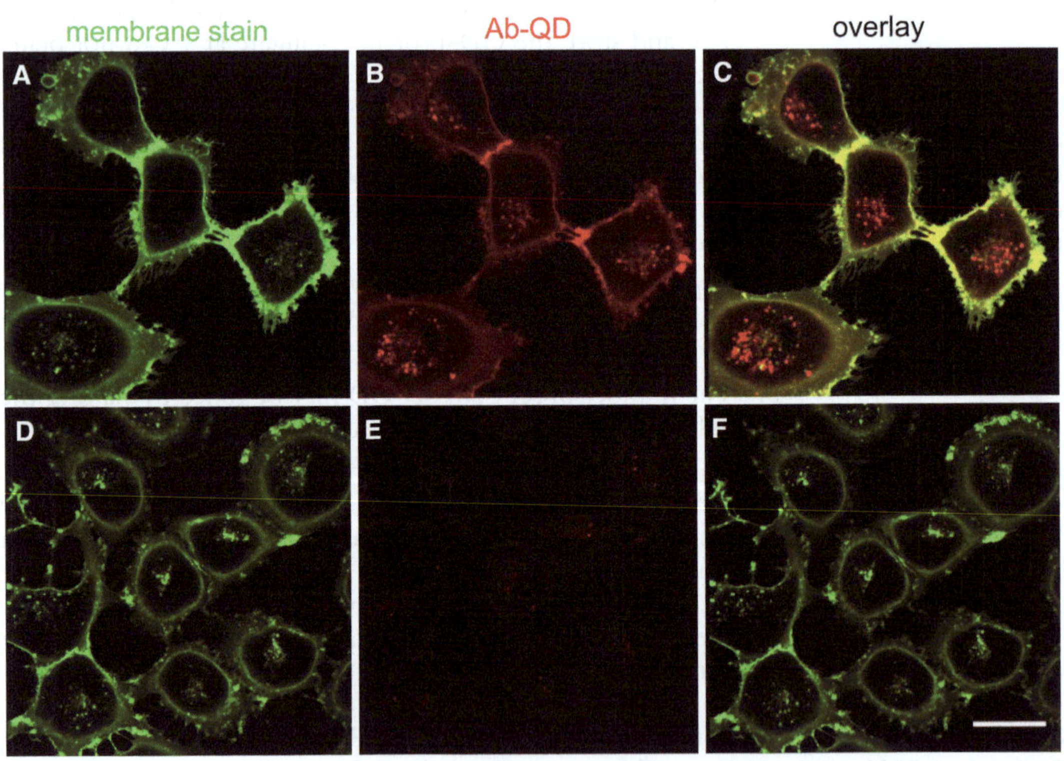

Fig. 5 Confocal images of live breast cancer MDA-MB-231 cells incubated with (**A–C**) scFv-QD conjugates (610 nm emission peak) or (**D–F**) SpyCatcher-QD conjugates without scFv Ab. (**A**, **D**) AlexaFluor 488-WGA fluorescence of the plasma membrane and endolysosomal pathway; (**B**, **E**) QD emission with or without surface anti-uPAR scFv. (**C**, **F**) Overlay of membrane and QD emission. QDs were excited at 405 nm and emission collected from 605 to 690 nm. AlexaFluor488 was excited at 488 nm and emission collected from 495–575 nm. Scale bar is 30 μm. Adapted from [17]

incubator, wash cells once with 500 μL of PBS + 10 mmol/L glucose.

4. If desired, cells may then be incubated with membrane stains or other counterstains in PBS + 10 mmol/L glucose at room temperature for 10 min, followed by washing three times with 500 μL of PBS + 10 mmol/L glucose.
5. Add fresh PBS + 10 mmol/L glucose, and image the cells (Fig. 5). Imaging conditions will depend on the microscope and cells being used (*see* **Note 11**). Generally, QDs are best excited by a 405 nm or 488 nm laser but may be excited anywhere up through the first exciton.

4 Notes

1. Excess surface ligand is important to prevent nanocrystals aggregation prior to polymer encapsulation. Addition of 1% (*v*/*v*) oleic acid, followed by brief sonication, works well.
2. PMAO of molecular weight 10 kDa or lower gives the best results for QDs smaller than 10 nm in diameter. Larger polymers may be used for larger nanoparticles.
3. Such as μ-slide 8 Well, IbiTreat, Polymer No. 1.5, and thickness of 180 μm.
4. It is important to ensure all organics are evaporated before adding the polymer. Slow, overnight evaporation under a gentle N_2 stream works well.
5. The ratio of unreactive PEG to protein-reactive amine can be varied as needed. Here, 1% of the PDEA is used with PEG3 as the excess surface ligand, but more or less can be incorporated as needed.
6. Other dialysis systems may be used to remove excess polymer provided they have 100 kDa molecular weight cut-offs.
7. More exact concentrations can be determined by using the height of the first exciton peak measured by UV-visible absorption [4].
8. Be sure to remove all thiol or TCEP reducing agents in the protein storage buffer using a desalting column (such as a Biospin-6 Desalting Column, Bio-Rad) or by centrifugal dialysis (such as Amicon Ultra-4, 3 kDa MWCO). Thiols will react with the activated disulfides needed for bioconjugation, and other reducing agents will reduce them.

9. QDs have a large absorption cross-section at 270 nm but no measurable emission, so the background subtraction is necessary for tryptophan emission from SpyCatcher, as well as any scattered light.
10. The American Type Culture Collection website (https://www.atcc.org/) has detailed cell culture protocols for many cell types, with details about cell growth media, cell surface requirements, and passaging protocols.
11. The optimal QD concentration for imaging will depend the brightness of the QDs, the cells being used, the receptors being targeted, and the particular type of microscopy. Live cell staining requires shorter incubation times than fixed cell staining. A number of conditions should be explored, and 10 nM is a good starting point for CdSe/CdS QDs, which have quantum yields >80%.

Acknowledgments

We thank Alex Powers, Sara Wichner, Cheryl Tajon, Lucas Thal, Behzad Rad, Jon Sack, Sandy Rosenthal, and Charly Craik for helpful discussion and experiments related to this work. This work was supported by National Institutes of Health award R01NS096317 (BEC). Work at the Molecular Foundry was supported by the Director, Office of Science, Office of Basic Energy Sciences, Division of Materials Sciences and Engineering of the U.S. Department of Energy under Contract No. DE-AC02-05CH11231.

References

1. Specht EA, Braselmann E, Palmer AE (2017) A critical and comparative review of fluorescent tools for live-cell imaging. Annu Rev Physiol 79:93–117. https://doi.org/10.1146/annurev-physiol-022516-034055
2. Basu K, Green EM, Cheng Y, Craik CS (2019) Why recombinant antibodies—benefits and applications. Curr Opin Biotechnol 60:153–158. https://doi.org/10.1016/j.copbio.2019.01.012
3. Han G, Mokari T, Ajo-Franklin C, Cohen BE (2008) Caged quantum dots. J Am Chem Soc 130:15811–15813. https://doi.org/10.1021/ja804948s
4. Wichner SM, Mann VR, Powers AS, Segal MA, Mir M, Bandaria JN, DeWitt MA, Darzacq X, Yildiz A, Cohen BE (2017) Covalent protein labeling and improved single-molecule optical properties of aqueous CdSe/CdS quantum dots. ACS Nano 11:6773–6781. https://doi.org/10.1021/acsnano.7b01470
5. Thal LB, Mann VR, Sprinzen D, McBride JR, Reid KR, Tomlinson ID, McMahon DG, Cohen BE, Rosenthal SJ (2020) Ligand-conjugated quantum dots for fast sub-diffraction protein tracking in acute brain slices. Biomater Sci 8:837–845. https://doi.org/10.1039/C9BM01629E
6. Qi X, Lee C, Ursprung B, Skripka A, Schuck PJ, Chan EM, Cohen BE (2024) Short-wave infrared upconverting nanoparticles. J Am Chem Soc 146:29292–29296. https://doi.org/10.1021/jacs.4c11181
7. Skripka A, Lee M, Qi X, Pan J-A, Yang H, Lee C, Schuck PJ, Cohen BE, Jaque D, Chan EM (2023) A generalized approach to photon avalanche upconversion in luminescent nanocrystals. Nano Lett 23:7100–7106. https://doi.org/10.1021/acs.nanolett.3c01955
8. Chen O, Zhao J, Chauhan VP, Cui J, Wong C, Harris DK, Wei H, Han H-S, Fukumura D, Jain RK, Bawendi MG (2013) Compact

high-quality CdSe–CdS core–shell nanocrystals with narrow emission linewidths and suppressed blinking. Nat Mater 12:445–451. https://doi.org/10.1038/nmat3539
9. Hildebrandt N, Spillmann CM, Algar WR, Pons T, Stewart MH, Oh E, Susumu K, Díaz SA, Delehanty JB, Medintz IL (2017) Energy transfer with semiconductor quantum dot bioconjugates: a versatile platform for biosensing, energy harvesting, and other developing applications. Chem Rev 117:536–711. https://doi.org/10.1021/acs.chemrev.6b00030
10. Caldwell MA, Albers AE, Levy SC, Pick TE, Cohen BE, Helms BA, Milliron DJ (2011) Driving oxygen coordinated ligand exchange at nanocrystal surfaces using trialkylsilylated chalcogenides. Chem Commun 47:556–558. https://doi.org/10.1039/C0CC02220A
11. Albers AE, Chan EM, McBride PM, Ajo-Franklin CM, Cohen BE, Helms BA (2012) Dual-emitting quantum dot/quantum rod-based nanothermometers with enhanced response and sensitivity in live cells. J Am Chem Soc 134:9565–9568. https://doi.org/10.1021/ja302290e
12. Saphire EO, Parren PW, Pantophlet R, Zwick MB, Morris GM, Rudd PM, Dwek RA, Stanfield RL, Burton DR, Wilson IA (2001) Crystal structure of a neutralizing human IGG against HIV-1: a template for vaccine design. Science 293:1155–1159. https://doi.org/10.1126/science.1061692
13. Kudirka R, Barfield RM, McFarland J, Albers AE, de Hart GW, Drake PM, Holder PG, Banas S, Jones LC, Garofalo AW, Rabuka D (2015) Generating site-specifically modified proteins via a versatile and stable nucleophilic carbon ligation. Chem Biol 22:293–298. https://doi.org/10.1016/j.chembiol.2014.11.019
14. Shen B-Q, Xu K, Liu L, Raab H, Bhakta S, Kenrick M, Parsons-Reponte KL, Tien J, Yu S-F, Mai E, Li D, Tibbitts J, Baudys J, Saad OM, Scales SJ, McDonald PJ, Hass PE, Eigenbrot C, Nguyen T, Solis WA, Fuji RN, Flagella KM, Patel D, Spencer SD, Khawli LA, Ebens A, Wong WL, Vandlen R, Kaur S, Sliwkowski MX, Scheller RH, Polakis P, Junutula JR (2012) Conjugation site modulates the in vivo stability and therapeutic activity of antibody-drug conjugates. Nat Biotechnol 30:184–189. https://doi.org/10.1038/nbt.2108
15. Szabó Á, Szendi-Szatmári T, Ujlaky-Nagy L, Rádi I, Vereb G, Szöllősi J, Nagy P (2018) The effect of fluorophore conjugation on antibody affinity and the photophysical properties of dyes. Biophys J 114:688–700. https://doi.org/10.1016/j.bpj.2017.12.011
16. Chudasama V, Maruani A, Caddick S (2016) Recent advances in the construction of antibody–drug conjugates. Nat Chem 8:114–119. https://doi.org/10.1038/nchem.2415
17. Pedroso CCS, Mann VR, Zuberbühler K, Bohn M-F, Yu J, Altoe V, Craik CS, Cohen BE (2021) Immunotargeting of nanocrystals by SpyCatcher conjugation of engineered antibodies. ACS Nano 15:18374–18384. https://doi.org/10.1021/acsnano.1c07856
18. Barat B, Sirk SJ, McCabe KE, Li J, Lepin EJ, Remenyi R, Koh AL, Olafsen T, Gambhir SS, Weiss S, Wu AM (2009) Cys-diabody quantum dot conjugates (immunoQdots) for cancer marker detection. Bioconjug Chem 20:1474–1481. https://doi.org/10.1021/bc800421f
19. Algar WR, Prasuhn DE, Stewart MH, Jennings TL, Blanco-Canosa JB, Dawson PE, Medintz IL (2011) The controlled display of biomolecules on nanoparticles: a challenge suited to bioorthogonal chemistry. Bioconjug Chem 22:825–858. https://doi.org/10.1021/bc200065z
20. Wegner KD, Jin Z, Lindén S, Jennings TL, Hildebrandt N (2013) Quantum-dot-based Förster resonance energy transfer immunoassay for sensitive clinical diagnostics of low-volume serum samples. ACS Nano 7:7411–7419. https://doi.org/10.1021/nn403253y
21. Francis JE, Mason D, Lévy R (2017) Evaluation of quantum dot conjugated antibodies for immunofluorescent labelling of cellular targets. Beilstein J Nanotechnol 8:1238–1249. https://doi.org/10.3762/bjnano.8.125
22. Mann VR, Powers AS, Tilley DC, Sack JT, Cohen BE (2018) Azide–alkyne click conjugation on quantum dots by selective copper coordination. ACS Nano 12:4469–4477. https://doi.org/10.1021/acsnano.8b00575
23. Agarwal P, Bertozzi CR (2015) Site-specific antibody-drug conjugates: the nexus of bioorthogonal chemistry, protein engineering, and drug development. Bioconjug Chem 26:176–192. https://doi.org/10.1021/bc5004982
24. Tilley DC, Eum KS, Fletcher-Taylor S, Austin DC, Dupré C, Patrón LA, Garcia RL, Lam K, Yarov-Yarovoy V, Cohen BE, Sack JT (2014) Chemoselective tarantula toxins report voltage activation of wild-type ion channels in live cells. Proc Natl Acad Sci USA 111:E4789–E4796. https://doi.org/10.1073/pnas.1406876111
25. Zakeri B, Fierer JO, Celik E, Chittock EC, Schwarz-Linek U, Moy VT, Howarth M (2012) Peptide tag forming a rapid covalent bond to a protein, through engineering a bacterial adhesin. Proc Natl Acad Sci USA 109:E690–E697. https://doi.org/10.1073/pnas.1115485109

26. Keeble AH, Turkki P, Stokes S, Khairil IN, Rahikainen R, Hytönen VP, Howarth M (2019) Approaching infinite affinity through engineering of peptide–protein interaction. Proc Natl Acad Sci USA 116:26523–26533. https://doi.org/10.1073/pnas.190965311
27. Mann VR, Manea F, Borys NJ, Ajo-Franklin CM, Cohen BE (2021) Controlled and stable patterning of diverse inorganic nanocrystals on crystalline two-dimensional protein arrays. Biochemistry 60:1063–1074. https://doi.org/10.1021/acs.biochem.1c00032
28. Charrier M, Li D, Mann V, Yun L, Jani S, Rad B, Cohen BE, Ashby PD, Ryan K, Ajo-Franklin CM (2018) Engineering the S-layer of *Caulobacter crescentus* as a foundation for stable, high-density, 2D living materials. ACS Synth Biol 8:181–190. https://doi.org/10.1021/acssynbio.8b00448
29. Tyler J, Ralston CY, Rad B (2023) Sneaking in SpyCatcher using cell penetrating peptides for in vivoimaging. Nanotechnology 34:425101.
30. Di Corato R, Quarta A, Piacenza P, Ragusa A, Figuerola A, Buonsanti R, Cingolani R, Manna L, Pellegrino T (2008) Water solubilization of hydrophobic nanocrystals by means of poly(maleic anhydride-alt-1-octadecene). J Mater Chem 18:1991–1996. https://doi.org/10.1039/B717801H
31. Pellegrino T, Manna L, Kudera S, Liedl T, Koktysh D, Rogach AL, Keller S, Rädler J, Natile G, Parak WJ (2004) Hydrophobic nanocrystals coated with an amphiphilic polymer shell: a general route to water soluble nanocrystals. Nano Lett 4:703–707. https://doi.org/10.1021/nl035172j
32. Tajon CA, Yang H, Tian B, Tian Y, Ercius P, Schuck PJ, Chan EM, Cohen BE (2018) Photostable and efficient upconverting nanocrystal-based chemical sensors. Opt Mater 84:345–353. https://doi.org/10.1016/j.optmat.2018.07.031
33. Fletcher-Taylor S, Thapa P, Sepela RJ, Kaakati R, Yarov-Yarovoy V, Sack JT, Cohen BE (2020) Distinguishing potassium channel resting state conformations in live cells with environment-sensitive fluorescence. ACS Chem Neurosci 11:2316–2326. https://doi.org/10.1021/acschemneuro.0c00276

Check for updates

Chapter 11

Conjugation of Quantum Dots to 3-Mercaptophenylboronic Acid for Analysis of Sialic Acids on Erythrocyte Surfaces

João Victor Araújo de Lima, Francisco Perazzo Tavares de Melo, Geyse Santos de Lima, Júlio Batista do Nascimento, Paulo Euzébio Cabral Filho and Adriana Fontes

Abstract

Phenylboronic acid derivatives (PBAs) are mimetic lectins that interact with 1,2- or 1,3-diols, groups commonly found in carbohydrates, including sialic acid (SA), a molecule involved in various biological processes, such as cell adhesion and signaling. SA is also responsible for the negative Zeta potential (ζ) of erythrocytes. What makes PBAs special, compared to other carbohydrate-recognizing molecules, is that their recognition properties can be modulated not only by environmental conditions, such as pH, but also by the addition of substituents to their structure. As a result, this organoborane has been applied for carbohydrate sensing and fundamental research in glycobiology. In addition, some PBAs, such as 3-mercaptophenylboronic acid (MPBA), contain thiol groups in their molecular structure, enabling their conjugation with nanostructures, such as quantum dots (QDs). QDs, when conjugated with MPBA, give rise to versatile fluorescent probes with advantageous properties capable of interacting with carbohydrates. Therefore, in this chapter, we present a protocol for the conjugation of hydrophilic QDs with MPBA via dative bonding and its application for studying the SA content in freshly collected and stored erythrocytes using fluorescence-based techniques. We hope that this chapter will be useful in guiding future work involving the application of QDs-MPBA conjugates in life sciences.

Keywords Carbohydrate, Fluorescence, Nanoprobe, Organoborane

1 Introduction

Among nanophotonic tools, quantum dots (QDs) are semiconductor fluorescent nanocrystals that possess unique physical and chemical properties. For example, QDs have wide absorption and narrow emission spectra, allowing QDs with different fluorescence wavelengths, in the same sample, to be excited by a single light source, while their fluorescent signals can still be individually resolved. They also have a high fluorescence intensity and resistance to photobleaching [1, 2]. QDs also possess a surface with functional groups that allow

Goreti Pereira et al. (eds.), *Quantum Dots*, Methods in Molecular Biology, vol. 3025,
https://doi.org/10.1007/978-1-0716-5210-7_11,

their conjugation to other (bio)molecules or nanoparticles. Owing to these properties, QDs have been widely applied in life science research as nanoprobes, for both *in vivo* and *in vitro* studies, as well as in optical and (photo)electrochemical sensors [3–5]. To this end, QDs have been conjugated with various (bio)molecules to enable specific interactions with target compounds in biological systems, enhancing sensitivity and versatility compared to traditional fluorescent probes, such as organic dyes [1, 6].

Phenylboronic acid derivatives (PBAs) are molecules that can be conjugated to the surfaces of QDs to develop fluorescent molecular recognition nanoprobes. These molecules have a boron atom with an empty *p* orbital, enabling reversible covalent binding to (bio) molecules containing diols, such as carbohydrates, resulting in the formation of five-or six-membered cyclic boronate esters [7]. Due to their reversible carbohydrate recognition properties, PBAs are also called mimetic lectins and can be widely applied as building blocks in synthetic chemistry, as well as for the identification and detection of (bio)molecules—even those that are structurally similar and present at low concentrations [8, 9].

These organoboranes can be modified by various substituents, including a thiol group at the 3-meta position. This structural modification results in 3-mercaptophenylboronic acid (MPBA), which is advantageous because its thiol group has a high affinity for metal and semiconductor surfaces, such as those of QDs, interacting with them through dative bonding. This makes the conjugation with these nanocrystals straightforward and fast [10, 11].

It is important to note that the addition of substituents can modify the electronic properties of PBAs by changing the charge distribution around the boron atom, which influences their pKa. Because the ionization of PBAs depends on both their pKa and the pH of the medium, variations in these parameters could lead to differences in their interaction with carbohydrates. These combined factors contribute to the complexity of PBA binding behavior [7].

Although some aspects of PBA reaction mechanisms remain to be clarified, these organoboranes can be applied in the construction of probes for recognizing carbohydrates important to the life sciences, such as sialic acid (SA) [12], with which they can form strong bonds. Studies by Lima et al. and Monteiro et al. have helped advance our understanding of the interaction between MPBA and carbohydrates, including SA [13, 14].

SA is a negatively charged structural carbohydrate present in glycoproteins on the surface of different cells and participates in processes, such as cell adhesion and signaling processes [15, 16]. The presence of SA in the erythrocyte membrane confers a negative Zeta potential (ζ) to the cells, keeping them separated and preventing their aggregation in blood vessels [15, 17]. Some diseases, such as hemoglobinopathies, could lead to a reduction in SA content, causing serious clinical conditions [18]. In addition, SA levels could be affected by erythrocyte storage conditions due to oxidative damage, which may impair the cells' ability to perform their functions effectively [19]. In this context, investigating changes in SA on the erythrocyte surface can help improve our understanding of their cell biology and contribute to optimizing erythrocyte storage conditions [13].

Thus, in this chapter, we present a protocol for preparing a conjugate that combines the optical properties of QDs with the recognition properties of MPBA (QDs-MPBA). We also discuss the application of the QDs-MPBA conjugate—developed by a simple and rapid method based on dative bonding—for analyzing SA content on the erythrocyte surfaces using fluorescence-based techniques. Thus, we hope that this chapter will stimulate future research on the SA profile in erythrocytes, as well as other cell types.

2 Materials

All reagent solutions must be prepared using ultrapure water (resistivity of 18.2 MΩ·cm) at room temperature (RT, approximately 25°C).

1. QDs stabilized/functionalized with MSA (*see* **Note 1**).
2. Mercaptosuccinic acid (MSA, 4.9% w/v) in ultrapure water.
3. Ultracentrifuge tubes with a 10-kDa filter cutoff.
4. MPBA solution (1.5 mg/mL) in ultrapure water (*see* **Note 2**).
5. Whole blood samples.
6. EDTA tubes.
7. Saline solution: NaCl 0.9% w/v in ultrapure water.

3 Methods

3.1 MPBA Conjugation with Quantum Dots

1. Adjust the pH of the suspension of MSA-CdTe QDs with the stabilizing agent itself: slowly add MSA solution dropwise into 1 mL of the QD suspension using a pipette with a maximum volume of 20 μL (*see* **Note 3**). Adjust the suspension pH to a range of 8.0–8.5 (*see* **Note 4**). Use a pH meter to monitor the pH during the addition of MSA.
2. Wash the pH-adjusted QD suspension to remove any remaining reagents of the synthesis using 10-kDa filter tubes, by centrifuging 4 times in 2-min cycles at 400×*g* (*see* **Note 5**). After each cycle, resuspend the QD suspension to its initial volume.
3. Add 400 μL of MPBA 1.5 mg/mL to 1 mL of the QD suspension (*see* **Note 6**).
4. Leave the suspension containing the QDs and MPBA under orbital stirring overnight at RT. Store the conjugate at 4°C for 7 days before application (*see* **Note 7**).

3.2 Preparation and Storage of Erythrocyte Suspension

1. Collect the whole blood samples in EDTA tubes (4 mL) and centrifuge them for 5 min at 455 × *g* to separate the erythrocytes from the other blood components (*see* **Note 8**).
2. Separate 500 µL of the erythrocytes by pipetting only from the pellet formed after centrifugation (*see* **Note 9**).
3. Add them to a microtube followed by the addition of 500 µL of saline solution (*see* **Note 10**). Homogenize and submit the cell suspension to three centrifugation cycles of 2 min at 260 ×*g* each. After every cycle, discard the supernatant and add 500 µL of saline solution to wash the erythrocytes.
4. Prepare a 1% (v/v) erythrocyte suspension by adding 10 µL of the washed erythrocyte suspension to 990 µL of saline solution.
5. To analyze stored blood samples, keep the whole blood collected in EDTA tubes at 4°C for the designated storage time. After that, follow the same steps shown in this section to prepare the 1% erythrocyte suspension (*see* **Note 11**).

3.3 Incubation of the QDs-MPBA Conjugate with Erythrocytes

1. Wash the conjugate to remove any MPBA molecules that have not interacted with QDs. For this, take an aliquot of the conjugate and centrifuge it using a 10 kDa cutoff tube for 3 min at 1120 ×*g*. After the centrifugation, discard the supernatant and resuspend the conjugate with saline solution to its initial volume.
2. Add the conjugate to the 1% erythrocyte suspension and keep it under orbital stirring at RT. Different proportions and incubation times could be used to establish the desired erythrocyte labeling, as described in the following sections.
3. Centrifuge the suspension at 260 ×*g* for 60 s, discard the supernatant, and resuspend the pellet using saline solution to its initial volume to remove any conjugate that did not bind to the SAs on the erythrocyte membrane.
4. To analyze stored blood samples, follow the same steps shown in Sect. 3.2.

3.4 Analysis by Fluorescence-Based Techniques

3.4.1 Fluorescence Microscopy

1. To confirm the conjugation of MPBA with QDs by fluorescence microscopy, incubate the conjugate with 1% erythrocyte suspension (see Sect. 3.3). We suggest starting by testing a cell:conjugate ratio of 5:1 (v/v), at RT, during a 5-min incubation. However, this ratio may depend on certain characteristics of the conjugate (*see* **Note 12**).
2. Wash the cell suspension after incubation by centrifugation at 260×*g* for 60 s and resuspend it to the initial volume using saline solution.

3. Take the suspension containing the conjugates and the cells to the fluorescence microscopy. It is important to define the fluorescence excitation and collection parameters based on the optical properties of the QDs used. An advantage of using QDs to develop the nanoprobe, is the wide absorption spectrum, which makes it easier to adapt the sample excitation wavelength.
4. Check if the interaction between conjugates and SA occurs through the MPBA, before incubating the erythrocytes with the conjugate. It is also possible to use cells treated with neuraminidase at 3 mU for 30 min at 37°C (*see* **Note 13**). Then, repeat the same steps above with the pretreated cells.
5. The images obtained by fluorescence microscopy will show intense labeling on the membranes of untreated erythrocytes (*see* **Note 14**). The cells treated with neuraminidase, in contrast, will rarely exhibit any labeling, since the QDs-MPBA conjugate does not interact with the cell surface because SA is removed during the washing steps prior to analyses, as illustrated in Fig. 1.

3.4.2 Analysis by Flow Cytometry

1. Incubate the 1% erythrocyte suspension, fresh or stored, with the QDs-MPBA conjugate. We suggest starting by testing a ratio of 2:1 (cell: conjugate—v/v) and an incubation time of 30 min at RT (*see* **Note 15**). Once again, these parameters depend on the characteristics of the conjugates (*see* **Note 12**).
2. Wash the system by centrifugation at 260 ×g for 60 s and resuspend it to its initial volume using saline solution.
3. Analyze the cells using a flow cytometer. We suggest the detection of at least 20,000 events in the gate. The parameters used must be selected based on the optical properties of the QDs in the study. The conjugation of QDs to MPBA is considered suitable, if the labeling of fresh erythrocytes approaches 100%.
4. To this analysis, start by selecting a gate within the FSC (Forward Scatter) *vs.* SSC (Side Scatter) plot. This step is essential for removing cellular debris from the analysis. Select the appropriate fluorescence excitation and detection parameters (e.g., filters), as well as the option to perform cell counting within the gate. Set the control at 1% in the gate positive in the histogram, corresponding to the autofluorescence of the erythrocytes.
5. Labeling by the QDs-MPBA conjugate can be sensitive to variations in SA concentration. To evaluate this, use erythrocyte samples stored for different periods. In general, the longer the storage time, the lower the fluorescent event count (y-axis), resulting in a reduced percentage of labeled cells and a broader histogram. These changes result from reduced interaction

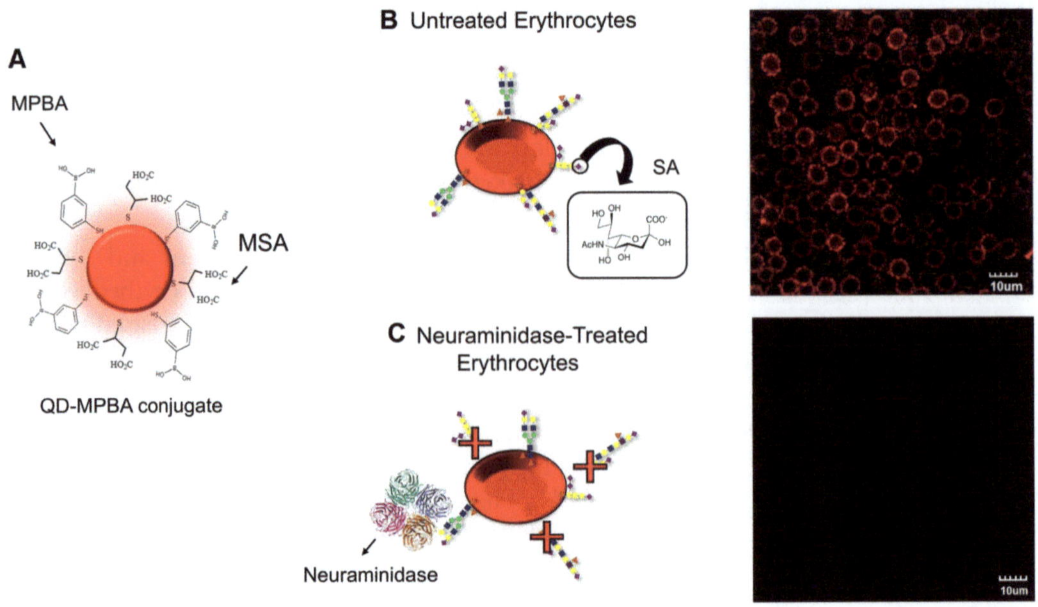

Fig. 1 In **A**, an illustration of the QD-MPBA conjugate and its components. In **B** and **C**, representative images of untreated and neuraminidase-treated erythrocytes after incubation with the conjugate, respectively (Scale bars: 20 μm). *QD* quantum dots, *MPBA* 3-mercaptophenylboronic acid, *MSA* mercaptosuccinic acid, *SA* sialic acid. Images in B and C were reproduced from Monteiro et al., Colloids and Surfaces B: Biointerfaces, © (2020), with permission from Elsevier [14]. Neuraminidase molecule is illustrated using the structure PDB ID: 2HTY [20]

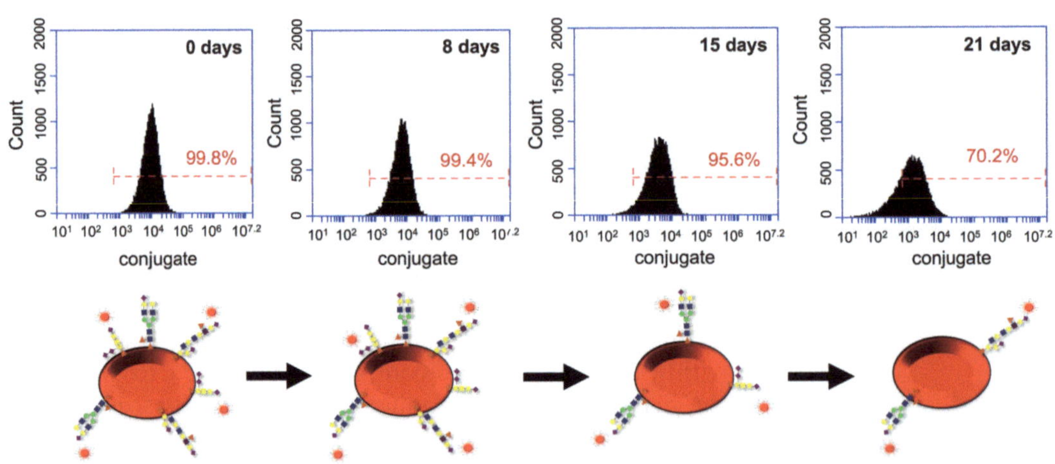

Fig. 2 Representative labeling of fresh (0 days) and stored erythrocytes (8, 15, 21 days) using the QDs-MPBA conjugate. The histograms were reproduced from Lima et al., Journal of Photochemistry and Photobiology A: Chemistry, © (2024), with permission from Elsevier [13]

between the conjugate and the cells due to lower SA content on the cell membrane (*see* **Note 16**), as illustrated in Fig. 2. We recommend performing the evaluations in at least triplicate.

4 Notes

1. It was used CdTe QDs stabilized/functionalized with MSA synthesized according to Lima et al. and Monteiro et al. [13, 14].
2. Use a microtube with a minimum capacity of 1.5 mL to weigh 1.5 mg of MPBA and add 1 mL of ultrapure water. Vigorously homogenize the solution using the pipette. Even using the pipette to homogenize, MPBA will not be completely diluted in water. For this reason, place the solution in an ultrasonic bath until the MPBA is completely solubilized. The average time is 15 min, and it is not necessary to heat the sample. However, it is still important to homogenize as much as possible with the pipette. This will make the subsequent step easier.
3. The pH of the suspension should be adjusted using the stabilizing/functionalizing agent employed in the synthesis—in our case, MSA. It helps maintain the colloidal stability and preserve the fluorescence of QDs.
4. The pH of the QD suspension should not be lower than 8.0 before adding MPBA. If this interval is not respected, the addition of MPBA will cause the suspension pH to drop below 7.0, making it difficult for the conjugate to interact with the SA. This is because MPBA has an acidic pH.
5. Washing is important to remove excess MSA and facilitate the interaction of MPBA to the QD surface, since MPBA contains a thiol group and binds to the QD via dative bonds. Another important aspect of washing is that the process can be more efficient if the nanosystem suspension is gently homogenized every two centrifugation steps using a 200 μL pipette.
6. We used QD concentrations ranging from 3 to 5 μM.
7. QDs-MPBA conjugate should be stored for 7 days at 4°C to ensure adequate interaction between the QDs and MPBA.
8. Studies involving erythrocytes require approval from an ethics committee. The ethical approvals for the studies presented herein are included in the original articles.
9. When centrifuged, a different pipetting method is required to collect the erythrocytes, because blood is viscous. It is important to use the reverse pipetting method to ensure that the correct volume of the erythrocytes is collected.

10. Saline is used because it is a medium that does not substantially damage the erythrocyte structure.
11. Although, we present a methodology for samples stored in EDTA tubes, the same procedures could be applied to studies using blood bags or alternative storage methods.
12. The ratio between cells and conjugates depend on the concentration of the QD suspension and the amount of MPBA present on their surface. For this reason, it is recommended to test different cell:conjugate ratios and incubation times to optimize labeling.
13. Treating the erythrocytes with neuraminidase before incubation with the QDs-MPBA conjugate will remove the SAs from the cell membrane. As a result, weak fluorescent labeling is expected under fluorescence microscopy.
14. In addition to erythrocytes, the protocol can be adapted for any cells that possess SA exposed on their membrane, whether they are blood cells or not, as reported in the study by Monteiro et al. [14].
15. The increased incubation time in flow cytometry, compared to microscopy, is associated with the lower amount of conjugate offered per cell.
16. The longer erythrocytes are stored, the more they undergo oxidative damage, which reduces the SA content in their membranes [19]. For this reason, erythrocytes stored for longer periods interact less with the QDs-MPBA conjugate, leading to decreased labeling, as the conjugate that does not interact with the cells is removed during centrifugation.

Acknowledgments

The authors are thankful to the National Council for Scientific and Technological Development (CNPq), the Coordination for the Improvement of Higher Education Personnel (CAPES), and the Foundation for Science and Technology of Pernambuco (FACEPE). This work is also linked to the National Institute of Photonics (INFo).

References

1. Lima JVA, Oliveira WF, Silva AR et al (2025) The brightness of lectins conjugated to quantum dots. Biophys Rev 17(2):419–434. https://doi.org/10.1007/s12551-025-01283-0
2. Borghei YS, Hosseinkhani S (2022) "Semiconductor quantum dots" in biomedical opportunities. J Lumin 243:118626. https://doi.org/10.1016/j.jlumin.2021.118626
3. Ribeiro JFF, Melo JRS, Santos C, de L et al (2023) Sensitive Zika biomarker detection assisted by quantum dot-modified electrochemical immunosensing platform. Colloids Surfaces

B 221:112984. https://doi.org/10.1016/j.colsurfb.2022.112984
4. Agarwal K, Rai H, Mondal S (2023) Quantum dots: an overview of synthesis, properties, and applications. Mater Res Express 10(6):062001. https://doi.org/10.1088/2053-1591/acda17
5. Yao J, Li L, Li P et al (2017) Quantum dots: from fluorescence to chemiluminescence, bioluminescence, electrochemiluminescence, and electrochemistry. Nanoscale 9(36):13364–13383. https://doi.org/10.1039/C7NR05233B
6. Henrique RBL, Lima JVA, Santos ALF et al (2024) Quantum dot-based conjugates: luminous nanotools for cancer research. Biochem Biophys Res Commun 741:151052. https://doi.org/10.1016/j.bbrc.2024.151052
7. Zhai W, Sun X, James TD et al (2015) Boronic acid-based carbohydrate sensing. Chem–Asian J 10:1836–1848. https://doi.org/10.1002/asia.201500444
8. Matsumoto A, Miyahara Y (2018) 'Boronolectin' based engineering as a versatile platform for biomedical applications. Sci Technol Adv Mater 19(1):18–30. https://doi.org/10.1080/14686996.2017.1411143
9. Lan T, Guo Q (2019) Phenylboronic acid-decorated polymeric nanomaterials for advanced bio-application. Nanotechnol Rev 8:548–561. https://doi.org/10.1515/ntrev-2019-0049
10. Henrique RBL, Santos ALF, Pereira MIA et al (2023) A fluorescent glyconanoprobe based on quantum dots and thiolated glucose: applications in monolayers and spheroids of cancer cells. Biochim Biophys Acta (BBA)-Gen Subj 1867(12):130474. https://doi.org/10.1016/j.bbagen.2023.130474
11. Chang L, He X, Chen L et al (2017) Mercaptophenylboronic acid-capped Mn-doped ZnS quantum dots for highly selective and sensitive fluorescence detection of glycoproteins. Sens Actuators B 243:72–77. https://doi.org/10.1016/j.snb.2016.11.121
12. Jangid AK, Kim K (2024) Phenylboronic acid-functionalized biomaterials for improved cancer immunotherapy via sialic acid targeting. Adv Colloid Interface Sci 333:103301. https://doi.org/10.1016/j.cis.2024.103301
13. Lima JVA, Melo FPT, Fontes A et al (2025) Probing the interaction of 3-mercaptophenylboronic acid with carbohydrates using quantum dots. J Photochem Photobiol A 464:116340. https://doi.org/10.1016/j.jphotochem.2025.116340
14. Monteiro CAP, Silva RC, Assis LG et al (2020) Quantum dots functionalized with 3-mercaptophenylboronic acids as novel nanoplatforms to evaluate sialic acid content on cell membranes. Colloids Surfaces B 193:111142. https://doi.org/10.1016/j.colsurfb.2020.111142
15. Doostkam A, Malekmakan L, Hosseinpour A et al (2022) Sialic acid: an attractive biomarker with promising biomedical applications. Asian Biomed Res Rev News 16(4):153–167. https://doi.org/10.2478/abm-2022-0020
16. Wei Y, Zheng L, Chen Y et al (2025) Analyzing and modifying sialic acid at the glycan end: systematic review of methods and functional outcomes. Trends Anal Chem 183:118083. https://doi.org/10.1016/j.trac.2024.118083
17. Zhu W, Zhou Y, Guo L et al (2024) Biological function of sialic acid and sialylation in human health and disease. Cell Death Discov 10:415. https://doi.org/10.1038/s41420-024-02180-3
18. Lima CN, Moura DS, Silva YSS et al (2020) Evaluating viscoelastic properties and membrane electrical charges of red blood cells with optical tweezers and cationic quantum dots—applications to β-thalassemia intermedia hemoglobinopathy. Colloids Surfaces B 186:110671. https://doi.org/10.1016/j.colsurfb.2019.110671
19. Huang YX, Tuo WW, Wang D et al (2016) Restoring the youth of aged red blood cells and extending their lifespan in circulation by remodelling membrane sialic acid. J Cell Mol Med 20:294-30. https://doi.org/10.1111/jcmm.12721
20. Russell RJ, Haire LF, Stevens DJ et al (2006) The structure of H5N1 avian influenza neuraminidase suggests new opportunities for drug design. Nature 443:45–49. https://doi.org/10.1038/nature05114

Chapter 12

Preparation of Ag_2Te Quantum Dots for Near-Infrared Fluorescence and Computed Tomography Imaging

Jiao-Jiao Ma, Ming-Xia Yu and Zhi-Quan Tian

Abstract

The combination of second near-infrared window (NIR-II) fluorescence imaging and high-resolution computed tomography (CT) provides complementary information for the diagnosis, progression, and prognosis of clinical conditions. Ag_2Te quantum dots (QDs) are promising CT/NIR-II fluorescence dual-modal imaging probes due to their high atomic number and excellent fluorescence performance. Herein, we describe the synthesis, characterization, and evaluation of the biocompatibility of Ag_2Te QDs in vitro and in vivo for CT/NIR-II fluorescence dual-modal imaging.

Keywords Second near-infrared fluorescence region, Ag_2Te quantum dots, Fluorescence imaging, CT imaging, Dual-modal imaging probes.

1 Introduction

Medical imaging is an indispensable tool for biomedical research and clinical diagnosis [1–3], providing essential information for clinical decision-making. It helps determine the extent, type, stage of disease, assess treatment efficacy, and detect tumor recurrence and metastasis. At present, commonly used clinical imaging diagnostic techniques rely on abnormal image manifestations, which primarily reflect gross morphological changes at the lesion site. However, these methods often lack histopathological information and specificity, making it difficult to comprehensively assess the disease from tissue structure to cellular and molecular processes. The advent of molecular imaging offers a promising solution to these challenges.

Currently, no single-mode imaging technology is perfect because clinical diagnostic methods with high sensitivity often suffer from low resolution, while those with high resolution tend

The authors J.-J. Ma, M.-X. Yu, and Z.-Q. Tian contributed equally.

Goreti Pereira et al. (eds.), *Quantum Dots*, Methods in Molecular Biology, vol. 3025,
https://doi.org/10.1007/978-1-0716-5210-7_12,

to have poor sensitivity. For example, computed tomography (CT) imaging is widely used in clinical diagnosis due to its advantages, including short scanning time, high-density contrast, high spatial resolution, and powerful image post-processing capability [4]. However, dynamic monitoring with CT remains challenging due to its low temporal resolution. In contrast, second near-infrared window (NIR-II) (1000–1700 nm) fluorescence imaging has gained significant attention in recent years for its low cost, high penetration depth, superior sensitivity, and ultrahigh temporal resolution [5, 6]. It has been successfully applied in real-time imaging and surgical navigation [7–14]. Combining CT imaging's superior spatial and density resolution with NIR-II fluorescence imaging's high sensitivity and temporal resolution is crucial for achieving accurate diagnosis. Multimode imaging is the development trend of molecular imaging, which integrates two or more imaging techniques, is a growing trend in molecular imaging that will significantly improve clinical research on diagnosis, prognosis, and intraoperative diseases monitoring [15].

The key to achieving dual-modality CT/NIR-II fluorescence imaging lies in developing probes with excellent X-ray attenuation properties and outstanding NIR-II fluorescence performance. Various imaging probes have been designed and applied in dual-mode imaging for numerous diseases [16, 17], including rare-earth nanoparticles [18–21], hybrid nanocomposites [22–26], and quantum dots (QDs) [27, 28]. Among these, I-VI Ag_2X (X=S, Se, Te) QDs demonstrate good biocompatibility, excellent optical properties, higher signal-to-noise ratio, greater penetration depth, and improved resolution in NIR-II fluorescence imaging. These characteristics make Ag_2X QDs highly promising as NIR-II fluorescence probes [29–37]. In addition, the high atomic numbers of Ag (Z=47) and Te (Z=52) endow Ag_2Te QDs with superior X-ray attenuation ability, which provides a strong theoretical basis for the application of Ag_2Te QDs in CT imaging [38].

Herein, we synthesized Ag_2Te QDs with a fluorescence emission peak at 1250 nm for CT/NIR-II fluorescence dual-modal imaging (Scheme 1, *see* **Note 1**) [27]. This approach avoids the complex multistep coupling reactions and addresses limitations, such as nephrotoxicity and the short residence time of iodinated molecules and their compounds in vivo [39–42]. We investigated the quantum dots' fluorescence stability, cytotoxicity, in vivo toxicity, biocompatibility, and preliminarily explored their fluorescence-CT multimodal imaging performance. Successful in vivo CT/NIR-II dual-mode imaging was achieved due to the quantum dots' stability, excellent X-ray attenuation, intense NIR-II fluorescence, small particle size, good biosafety, and high atomic number.

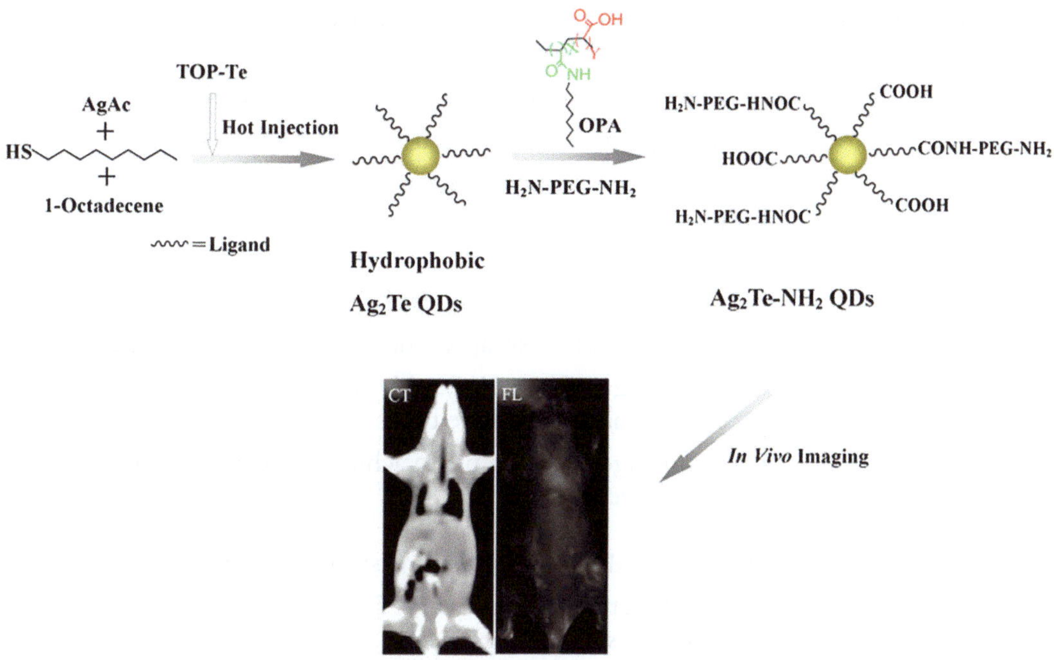

Scheme 1 Schematic diagram of the Ag_2Te QDs used as contrast agents for NIR-II fluorescence and CT Imaging. Reprinted from [27] with permission

2 Materials

2.1 Hydrophilic Ag_2Te QDs

1. Tetramethylammonium hydroxide aqueous solution, 25%.
2. Hydrochloric acid (HCl) aqueous solution, 1 mol/L.
3. Borate saline buffer, 50 mmol/L, pH = 9.0.
4. 10 mL separation column (30 cm × 10 mm), 10,000–600,000 Da (where it was used a Superdex 200).
5. Borate saline buffer, 20 mmol/L, pH = 7.4.
6. Borate saline buffer, 50 mmol/L, pH = 8.4.
7. Desalting column, 1 mL NAP-10 (where it was used a Sephadex G-25 DNA Grade).
8. 50× Tris–acetate–EDTA (TAE) solution: 0.484 mg Tris acid, 0.74 mg EDTA, and 1.4 mL of acetic acid were thoroughly mixed in deionized water to a constant volume of 1 L.
9. The silicon wafers were immersed in the pretreatment solution for 12 h, 30% hydrogen peroxide: 98% concentrated sulfuric acid solution at a 3:7 (v/v) ratio.
10. NH_2-PEG-NH_2 (MW2 kDa) solutions: 1.5 mg NH2-PEG-NH2, 0.5 mL BR (pH 7.4, 20 mmol/L).
11. Indocyanine green (ICG, 90%), ICG was dissolved in dimethyl sulfoxide (DMSO).

2.2 Characterization

1. Transmission electron microscope (TEM) with energy dispersive X-ray (EDX).
2. X-ray powder diffractometer (XRD, Cu Kα radiation, 0.5°/min).
3. X-ray photoelectron spectrometer (XPS).
4. UV-Vis spectrophotometer.
5. Fluorescence spectrophotometer (InGaAs detector, between 700 and 1600 nm).
6. Dynamic light scattering and Zeta potential equipment.
7. Gel electrophoresis apparatus, 1% (w/v) agarose gel.
8. Centrifuge (TG16-WS): able to achieve 17,800 ×*g* and hold 1.5 mL microcentrifuge tubes.
9. Centrifuge (5430): able to achieve 30,130 ×*g* and hold 50 mL centrifuge tubes.

2.3 In Vitro and In Vivo Studies

1. MDA-MB-231 human breast cancer cells.
2. 5-Dimethylthiazol-2-yl-2, 5-diphenyl tetrazolium bromide (MTT) solution, 5 mg/mL.
3. Human whole blood.
4. Centrifuge: able to achieve 12,000 ×*g* and hold 1.7 mL microcentrifuge tubes.
5. Centrifuge: able to achieve 3000 ×*g* and hold 50 mL centrifuge tubes.
6. Female Balb/c mice (8-week-old).
7. Phosphate buffered saline, 20 mmol/L, pH = 7.4.
8. Blood collection tubes containing anticoagulants.
9. Vet analyzer.
10. Isoflurane, 2%.
11. NIR-II imaging system equipped with an InGaAs camera (1000 nm long-pass filter, excitation at 808 nm).

3 Methods

3.1 Hydrophobic Ag_2Te QDs

The procedure for preparing hydrophobic NIR-II Ag_2Te QDs may be found in [43], which is outlined below.

1. Add 76.6 mg Te powder and 9 mL tri-*n*-octylphosphine (TOP, 90%) to 100 mL round-bottom flask, and preserve it in an inert atmosphere (*see* **Note 2**).
2. Raise the temperature to 110°C, and maintain this temperature for 30 min.

3. Obtain a pale-yellow transparent TOP-Te precursor solution.
4. Cool the solution in Step 3 naturally to room temperature (20–25°C) for use.
5. Add 5 mL of 1-octadecene (ODE, 90%) to another 100 mL round-bottom flask at 50°C under an inert atmosphere for 30 min (*see* **Note 3**).
6. Mix 260 μL of 1-octanethiol (OT, ≥98.5%) and 33.4 mg of silver acetate (AgAc) to the flask in Step 5.
7. Increase the temperature of the mixture gradually to 170°C.
8. Quickly inject 0.3 mL of TOP-Te solution (*see* **Note 4**).
9. The temperature was decreased from 170 to 120°C and the reaction was constant at 120°C for 60 min.
10. Wash the products with ethanol (the v/v ratio of QDs suspension and ethanol is 1/9) and centrifuge for 3 min (1700 ×*g*), and finally disperse them in hexane or tetrachlorethylene (*see* **Note 5**).

3.2 Hydrophilic Ag_2Te QDs

The procedure for preparing Hydrophilic Ag_2Te QDs follows a similar method to our previous report on water-dispersible Ag_2Se QDs [44]. Using this approach, Ag_2Te–COOH and Ag_2Te–NH_2 QDs are successfully synthesized.

3.2.1 Preparation of Octamine-Modified Polyacrylic Acid (OPA)

1. Place 5.32 g of polyacrylic acid 2000 (PAA, MW: 1800) and 5.7 g *N*-(3-dimethylaminopropyl)-*N'*-ethylcarbodiimide hydrochloride (EDC) into a 100 mL round-bottom flask.
2. Add 50 mL of *N,N*-dimethylformamide (DMF) to dissolve the mixture (*see* **Note 6**).
3. Add 3.2 mL of octylamine (99%) dropwise into the reaction flask and then stir it overnight. At the end of the reaction, the solution should be a yellowish clear liquid.
4. Place the reaction solution in Step 3 on the rotary evaporator at 90°C vacuum to remove DMF.
5. Dissolve the adhesive substance in Step 4 with 30 mL acetone, and wash the precipitation with ultrapure water, and repeat it three times (*see* **Note** 7).
6. Redissolve the precipitate in Step 5 in 80 mL of ethyl acetate at 50°C and transfer it to a separating funnel, add 40 mL of tetramethylammonium hydroxide aqueous solution (25%), and then stratify it after full oscillation (*see* **Note 8**).
7. Add 1 M HCl to the lower solution obtained in Step 6 (the v/v ratio of lower solution obtained in Step 6 and HCl is 1/3) and centrifuge for 10 min (12,000 ×*g*), repeat the washing with ultrapure water.

8. Dissolve the precipitate obtained in Step 7 with a small amount of ethanol (5–10 mL), and then triploid 1 mol/L HCl was added to the solution. Separate the precipitate by centrifugation again with ultrapure water. Repeat this procedure at least five times.
9. Dissolve the precipitate obtained in Step 8 with ethanol (5–10 mL), and remove the ethanol by a rotary evaporator to obtain a white or yellowish solid powder (OPA), which is collected (*see* **Note 9**).

3.2.2 Ag_2Te–COOH QDs

Ag_2Te QDs (prepared in Sect. 3.1) and OPA (prepared in Sect. 3.2.1) react in chloroform according to a certain mass ratio (the m/m ratio of hydrophobic QDs and OPA is 1/1.5).

1. Dissolve 15 mg OPA in 2 mL chloroform.
2. Add 10 mg hydrophobic Ag_2Te QDs drop by drop to the solution in Step 1.
3. Dry by rotary evaporation the suspension in Step 2.
4. Dissolve the dried solid in 4 mL borate saline buffer (pH = 9.0, 50 mmol/L) and purify it by gel filtration on polypropylene columns to get Ag_2Te–OPA QDs.

3.2.3 Ag_2Te–NH_2 QDs

1. 10 mg Ag_2Te–COOH QDs were dispersed in borate saline buffer (pH = 7.4, 20 mmol/L), and then activate it by adding 15 mg EDC.
2. Add 15 mg NH_2–PEG–NH_2 (MW 2 kDa) into the suspension of activated Ag_2Te–COOH QDs (m/m ratio of Ag_2Te–COOH QDs, EDC, and NH_2–PEG–NH_2 is $m_{QDs}:m_{EDC}:m_{NH2\text{-}PEG\text{-}NH2} = 1:1.5:1.5$).
3. React it at room temperature for 2 h, then the reaction was terminated with BR buffer pH = 8.4 (*see* **Note 10**).
4. Concentrate the solution of Step 3 by centrifugation for 5 min (12,000 × *g*).
5. The solution in Step 4 was purified by gel filtration on NAP-10 columns to get Ag_2Te–NH_2 QDs.

3.3 Characterization

3.3.1 Transmission Electron Microscopy (TEM) and Energy Dispersive X-ray (EDX) Spectroscopy

TEM is used to confirm the morphology, crystal structure, and size of Ag_2Te QDs. EDX is used to obtain the information of chemical composition and elemental content of Ag_2Te QDs.

1. Disperse the hydrophobic Ag_2Te QDs in an appropriate concentration (0.2 mg/mL) with toluene and dropped onto an ultrathin carbon film.

2. After the solvent evaporates, observe the QDs using a transmission electron microscope (Fig. 1A).
3. Disperse the hydrophilic Ag_2Te QDs in ultrapure water (0.2 mg/mL), drop it onto an ultrathin carbon film, and allow it to air dry overnight.
4. Observe the QDs using a transmission electron microscope (Fig. 1B).

3.3.2 The X-ray Diffraction (XRD)

XRD was employed to determine the crystal structure, crystal parameters, and crystal orientation of the prepared nanomaterials (Fig. 2).

1. Disperse the purified hydrophobic Ag_2Te QDs in toluene (3 mg/mL) and drop them onto the center of a slide.
2. After the solvent evaporates, place the slide with samples on the sample holder of a X-ray diffraction instrument for scanning. The scanning range was 20–70°, with a scanning speed of 0.5°/min.

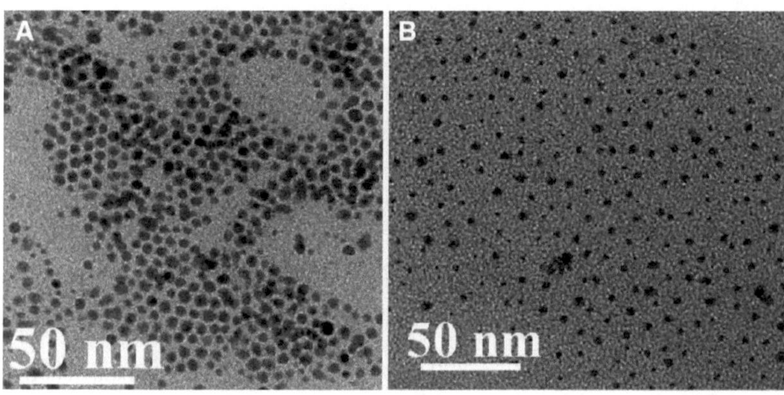

Fig. 1 TEM images of hydrophobic (**A**) and hydrophilic (**B**) Ag_2Te QDs

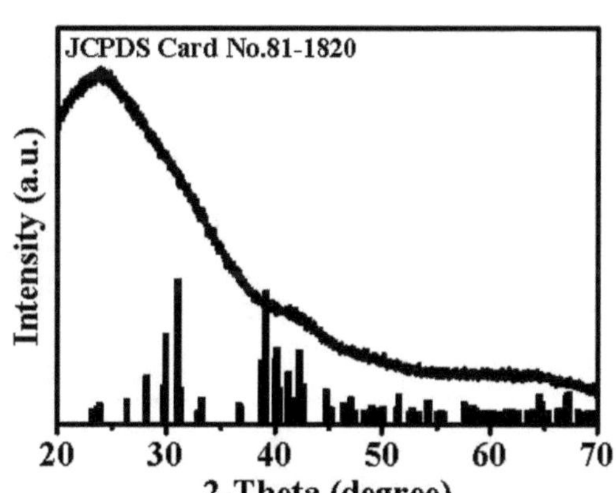

Fig. 2 XRD pattern of hydrophobic Ag_2Te QDs

3.3.3 X-ray Photoelectron Spectroscopy (XPS)

XPS is used to analyze the elemental composition, chemical state, and electronic structure of the material surface (Fig. 3, *see* **Note 1**).

1. Disperse the purified hydrophobic Ag_2Te QDs in toluene (3 mg/mL) and drop them onto a silicon wafer substrate (silicon wafers were immersed in a mixed solution of hydrogen peroxide and sulfuric acid for 12 h).
2. Analyze the samples using a X-ray photoelectron spectrometer.

3.3.4 UV-Vis Absorption Spectrum and Fluorescence Emission Spectroscopies

UV-Vis absorption spectroscopy is used to measure the absorption characteristics of the nanoparticles in the UV-visible range, including the position and intensity of absorption peaks, to characterize the particle morphology, size, and composition (Fig. 4).

The fluorescence emission spectra are used to analyze the fluorescence emission characteristics of the nanomaterials, including the position and intensity of emission peaks, as well as the fluorescence properties of the particles and the surface modification effects.

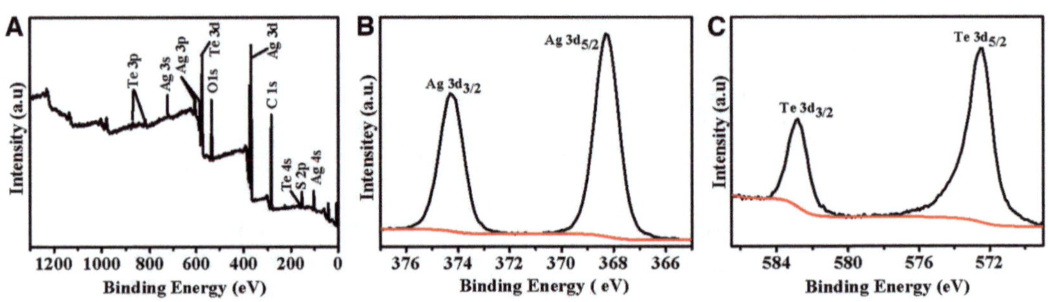

Fig. 3 XPS spectra of hydrophobic Ag2Te QDs: XPS survey spectrum (**A**), high-resolution XPS spectra of Ag 3d (**B**) and Te 3d (**C**)

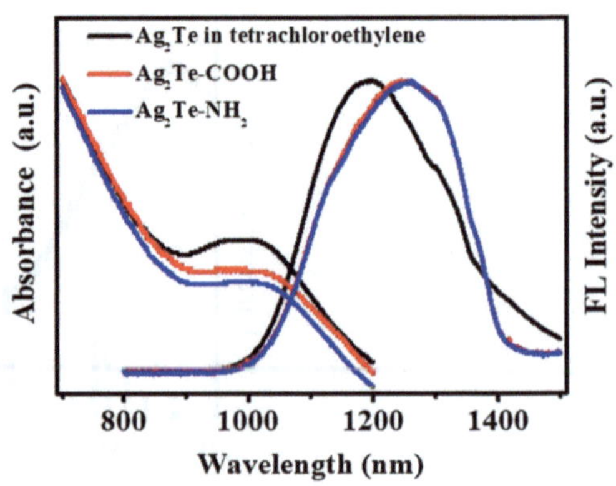

Fig. 4 UV-Vis-NIR and fluorescence spectra of hydrophobic Ag_2Te and hydrophilic Ag_2Te QDs.

1. Disperse the hydrophobic Ag_2Te QDs to the appropriate concentration with tetrachloroethylene (500 μL, 0.3 mg/mL).
2. Disperse the hydrophilic Ag_2Te QDs samples with water (500 μL, 0.3 mg/mL).
3. Place the samples in a quartz cuvette and scan it in a spectrophotometer from 700 to 1600 nm, for absorption measurements.
4. For fluorescence measurements, place the samples into a quartz cuvette and measure it in a fluorescence spectrophotometer. The excitation wavelength should be set at 400 nm, excitation and emission slits of 10 and 15 nm, respectively. The emission spectra are scanned from 800 to 1600 nm.

3.3.5 Dynamic Light Scattering (DLS)

DLS is used to measure the size distribution and dynamic behavior of the particles, such as diffusion coefficient and aggregation state of nanomaterials (Fig. 5A, *see* **Note 1**).

Zeta potential measurements are used to assess the particle surface charge and stability of the nanomaterials.

1. Disperse the hydrophilic Ag_2Te QDs to an appropriate concentration in water (500 μL 0.3 mg/mL).
2. Add it to the sample cell and analyze it using a nanoparticle size and surface potential analyzer, with gain values in the range of 7 to 11.

3.3.6 Agarose Gel Electrophoresis

Agarose gel electrophoresis is useful for confirming successful ligand exchange with –COOH and $–NH_2$ groups, as the electrophoretic mobility of the QDs depends on their charge and size. –COOH QDs are negatively charged, while $–NH_2$ QDs are approximately neutral (Fig. 5B, *see* **Note 1**).

1. To prepare the gel, mix 150 mg of agarose and 15 mL of 0.5 × TAE solution and heat it in a microwave to 90–100°C for 1 min to completely dissolve the agarose.

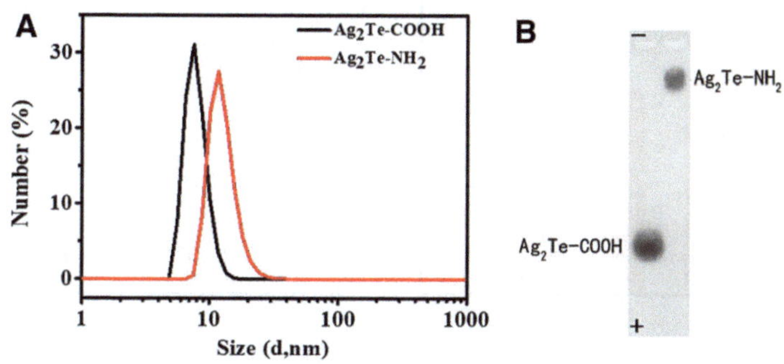

Fig. 5 Hydrodynamic diameters (**A**) and agarose gel electrophoresis (**B**) of Ag_2Te–COOH QDs and $Ag_2Te–NH_2$ QDs. Reprinted from [27] with permission

2. Pour the hot solution quickly into the mold and allow it to cool at room temperature for 20 min until it solidifies into a gel.
3. Then, place the cooled agarose gel in the gel electrophoresis tank.
4. Add 100 μL samples to the well, turn the power supply on, and run the gel in constant voltage mode 120 mV for 30 min.
5. Afterward, disconnect the power, and remove the agarose gel for further use.

3.3.7 Stability of Ag_2Te–NH_2 QDs

1. Disperse the Ag_2Te–NH_2 QDs in BR buffer (pH 8.4) and store it at 4°C.
2. Measure the fluorescence emission spectra of the sample every 7 days to investigate the stability of Ag_2Te–NH_2 QDs (Fig. 6A).
3. To study their photostability, continuously irradiate Ag_2Te QDs and the fluorescent dye indocyanine green (ICG, 90%, used as a control) by 3.5 W/cm² of laser (808 nm) for 1 h.
4. Record their fluorescence spectra every 5 min (Fig. 6B)

3.4 In Vitro and In Vivo Studies

3.4.1 In Vitro Cytotoxicity and Hemolysis Assay

Cytotoxicity and hemolysis assays are performed to evaluate the biotoxicity and biocompatibility of Ag_2Te QDs in vitro.

1. Seed logarithmic MDA-MB-231 cells in a 96-well plate (10,000 cells/well) and incubate with various concentrations (0, 1.5, 3, 6, 12, 25, 50, 100, 200 μg/mL) of 100 μL Ag_2Te–NH_2 QDs for 48 h.
2. After incubation, add 20 μL of MTT solution, and incubate the cells for 4 h.
3. Subsequently, add 100 μL of DMSO to replace the DMEM (Dulbecco's Modified Eagle's medium), and gently shake the mixture for 10 min.

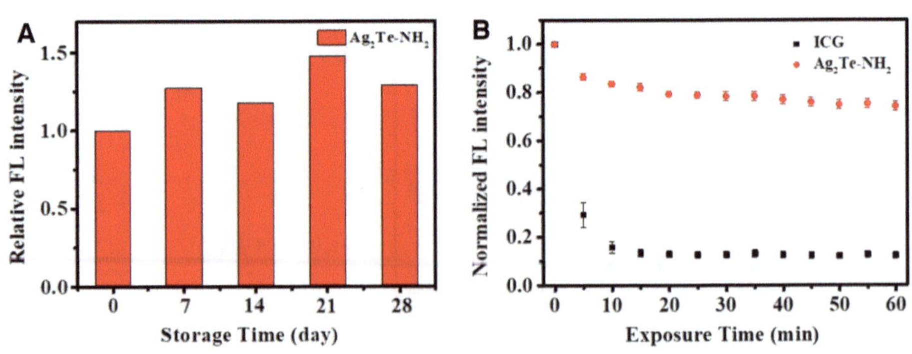

Fig. 6 The fluorescence intensities of Ag_2Te–NH_2 QDs stored in BR buffer (pH 8.4) at 4°C for 1 month (**A**) and photostability of Ag_2Te–NH_2 QDs and ICG under continuous irradiation of 3.5 W cm^{-2} 808 nm laser (**B**)

4. Measure the absorbance at 570 nm using a microplate spectrophotometer.
5. For the hemolysis assay, use human whole blood (*see* **Note 11**).
6. Add different concentrations of $Ag_2Te–NH_2$ QDs (3, 6, 12, 25, 50, 100, 200 μg/mL) to 100 μL human whole blood.
7. Incubate the mixtures at 37°C for 4 h.
8. After incubation, collect the serum by centrifugation for 10 min at 1300 ×*g*, followed by a second centrifugation for 15 min at 14,000 ×*g* to remove unreacted $Ag_2Te–NH_2$ QDs and the supernatants.
9. Record the absorbance at 570 nm using a spectrophotometer.
10. Calculate the hemolysis rate using the following formula:

$$\text{Hemolysis rate } (\%) = (A - A-)/(A+ - A-) \times 100\%$$

Where A, $A-$, and $A+$ are absorbance values (at 570 nm) of sample, negative control group, and positive control group, respectively.

3.4.2 In Vivo Biocompatibility and Histological Analysis

The in vivo biotoxicity and biocompatibility of Ag_2Te QDs are evaluated by monitoring changes in body weight and blood indexes of mice at different time points after the injection of $Ag_2Te–NH_2$ QDs. The potential toxicity of Ag_2Te QDs is further assessed by histopathological examination (*see* **Note 1**).

Female Balb/c mice (8-week-old) after purchase are housed in a specific-pathogen-free (SPF) environment for further use.

1. Divide 15 mice randomly and equally into five groups for time points of 6 h, 1 day, 3 days, 7 days, and 14 days.
2. Inject mice into the tail-vein in the experimental and control groups with $Ag_2Te–NH_2$ QDs (100 μL 3 mg/mL) and PBS (100 μL), respectively.
3. Collect, at each time point, 300 μL of fresh whole blood from the mice using blood collection tubes containing anticoagulants.
4. Analyze the samples using a Vet analyzer.

3.4.3 In Vivo NIR-II Fluorescence and CT Imaging

1. The mice were anesthetized with 2% isoflurane in 2 L/min O_2 gas, and laid on a warmer to maintain the body temperature, and scan them as baseline for in vivo imaging.
2. Then, inject tail-vein 100 μL $Ag_2Te–NH_2$ QDs (3 mg/mL), with an emission peak at 1250 nm, into the mice for in vivo NIR-II fluorescence and CT imaging (Fig. 7, *see* **Note 1**).

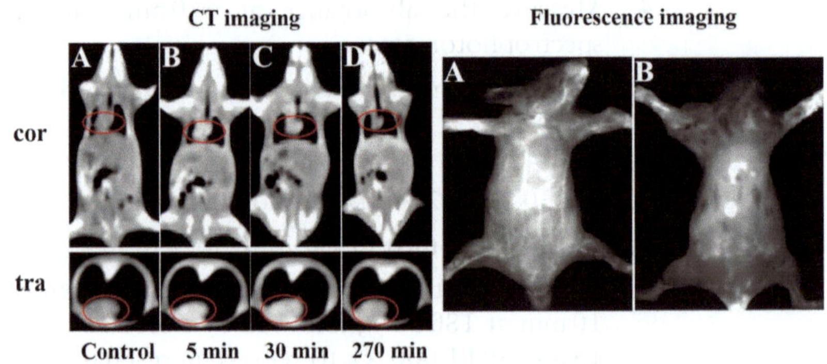

Fig. 7 In vivo CT imaging of mice injected with $Ag_2Te–NH_2$ QDs (with red circles indicating heart strength): (**A**) control group and at (**B**) 5 min, (**C**) 30 min, and (**D**) 4.5 h postinjection (cor: coronal position, tra: transverse position) (Reprinted from [27] with permission.). In vivo NIR-II fluorescence imaging of mice injected with $Ag_2Te–NH_2$ QDs at different position (**A**) supine and (**B**) prone (excited with a 4 W/cm^2 808 nm laser at an exposure time of 200 ms)

4 Notes

1. Reprinted with permission from Ag_2Te quantum dots as contrast agents for near-infrared fluorescence and computed tomography imaging. Copyright 2020 American chemical society.
2. The Te powder and TOP should be stored in the anaerobic glove box, and the weighing should be completed in the glove box. The samples taken should be temporarily stored in the transition warehouse before being taken out.
3. Hydrophobic Ag_2Te QDs were prepared in an inert atmosphere.
4. Te precursor should be injected quickly.
5. The supernatant was collected as the resulting Ag_2Te QDs.
6. Stir and dissolve at room temperature.
7. After centrifugation in the centrifuge, the liquid was poured away to retain the solid, and the wash was repeated three times.
8. Collect the yellow liquid part (upper layer).
9. Store in desiccator for use.
10. The reaction was carried out for 2 h at room temperature (20–25°C) with shaking, and the borate saline buffer concentration was 50 mmol/L at pH = 8.4.
11. All the samples were collected with informed consent from the participants, and treatment in accordance with the approved Ethics Committee.

References

1. Kairdolf BA, Qian X, Nie S (2017) Bioconjugated nanoparticles for biosensing, in vivo imaging, and medical diagnostics. Anal Chem 89(2):1015–1031. https://doi.org/10.1021/acs.analchem.6b04873
2. Jaehong, Key, James et al (2014) Nanoparticles for multimodal in vivo imaging in nanomedicine. Int J Nanomed 9(1):711–726. https://doi.org/10.2147/IJN.S53717
3. James ML, Gambhir SS (2012) A molecular imaging primer: modalities, imaging agents, and applications. Physiol Rev 92(2):897–965. https://doi.org/10.1152/physrev.00049.2010
4. Schenkman L (2011) Second thoughts about CT imaging. Science 331:1002–1004. https://doi.org/10.1126/science.331.6020.100
5. Li C, Chen G, Zhang Y, Wu F, Wang Q (2020) Advanced fluorescence imaging technology in the near-infrared-II window for biomedical applications. J Am Chem Soc 142:14789–14804. https://doi.org/10.1021/jacs.0c07022
6. He S, Song J, Qu J, Cheng Z (2018) Crucial breakthrough of second near-infrared biological window fluorophores: design and synthesis toward multimodal imaging and theranostics. Chem Soc Rev 47:4258–4278. https://doi.org/10.1039/C8CS00234G
7. Wang W, Kong Y, Jiang J, Xie Q, Huang Y, Li G, Wu D, Zheng H, Gao M, Xu S et al (2020) Engineering the protein corona structure on gold nanoclusters enables red-shifted emissions in the second near-infrared window for gastrointestinal imaging. Angew Chem Int Ed 59:22431–22435. https://doi.org/10.1002/anie.202010089
8. Smith AM, Mancini MC, Nie S (2009) Bioimaging: second window for in vivo imaging. Nat Nanotechnol 4:710–711. https://doi.org/10.1038/nnano.2009.326
9. Fan X, Xia Q, Zhang Y, Li Y, Feng Z, Zhou J, Qi J, Tang BZ, Qian J, Lin H (2021) Aggregation-induced emission (AIE) nanoparticles-assisted NIR-II fluorescence imaging-guided diagnosis and surgery for inflammatory bowel disease (IBD). Adv Healthc Mater 10:2101043. https://doi.org/10.1002/adhm.202101043
10. Kenry, Duan Y, Liu B (2018) Recent advances of optical imaging in the second near-infrared window. Adv Mater 30:1802394. https://doi.org/10.1002/adma.201802394
11. Yao C, Chen Y, Zhao M, Wang S, Wu B, Yang Y, Yin D, Yu P, Zhang H, Zhang F (2022) A bright, renal-clearable NIR-II brush macromolecular probe with long blood circulation time for kidney disease bioimaging. Angew Chem Int Ed 61:e202114273. https://doi.org/10.1002/anie.202114273
12. Huang JG, Xie C, Zhang XD, Jiang YY, Li JC, Fan QL, Pu KY (2019) Renal-clearable molecular semiconductor for second near-infrared fluorescence imaging of kidney dysfunction. Angew Chem Int Ed 58:15120–15127. https://doi.org/10.1002/anie.201909560
13. Liu HL, Hong GS, Luo ZT, Chen JC, Chang JL, Gong M, He H, Yang J, Yuan X, Li LL et al (2019) Atomic-precision gold clusters for NIR-II imaging. Adv Mater 31:1901015. https://doi.org/10.1002/adma.201901015
14. Jing L, Yang C, Zhang P, Zeng J, Li Z, Gao M (2020) Nanoparticles weaponized with built-in functions for imaging guided cancer therapy. View 1:e19. https://doi.org/10.1002/viw2.19
15. Crozals GD, Bonnet R, Farre C, Chaix C (2016) Nanoparticles with multiple properties for biomedical applications: a strategic guide. Nano Today 11:435–463. https://doi.org/10.1016/j.nantod.2016.07.002
16. Wen L, Chen L, Zheng S, Zeng J, Duan G, Wang Y, Wang G, Chai Z, Li Z, Gao M (2016) Ultrasmall biocompatible WO_{3-x} nanodots for multi-modality imaging and combined therapy of cancers. Adv Mater 28:5072–5079. https://doi.org/10.1002/adma.201506428
17. Wei Z, Duan G, Huang B, Qiu S, Zhou D, Zeng J, Cui J, Hu C, Wang X, Wen L et al (2021) Rapidly liver-clearable rare earth core-shell nanoprobe for dual-modal breast cancer imaging in the second near-infrared window. J Nanobiotechnol 19:369. https://doi.org/10.1186/s12951-021-01112-y
18. Zhong YT, Dai HJ (2020) A mini-review on rare-earth down conversion nanoparticles for NIR-II imaging of biological systems. Nano Res 13:1281–1294. https://doi.org/10.1007/s12274-020-2721-0
19. Dai Y, Yang D, Yu D, Cao C, Wang Q, Xie S, Shen L, Feng W, Li F (2017) Mussel-inspired polydopamine-coated lanthanide nanoparticles for NIR-II/CT dual imaging and photothermal therapy. ACS Appl Mater Interfaces 9:26674–26683. https://doi.org/10.1021/acsami.7b06109
20. Kang N, Liu Y, Zhou Y, Wang D, Chen C, Ye S, Nie L, Ren L (2016) Phase and size control of core-shell upconversion nanocrystals light up deep dual luminescence imaging and CT in vivo. Adv Healthc Mater 5:1356–1363. https://doi.org/10.1002/adhm.201600159
21. Xing H, Bu W, Ren Q, Zheng X, Li M, Zhang S, Qu H, Wang Z, Hua Y, hao K et al (2012) A

$NaYbF_4$: Tm^{3+} nanoprobe for CT and NIR-to-NIR fluorescent bimodal imaging. Biomaterials 33:5384–5393. https://doi.org/10.1016/j.biomaterials.2012.04.002
22. Nakamura M, Awaad A, Hayashi K, Ochiai K, Ishimura K (2012) Thiol-organosilica particles internally functionalized with propidium iodide as a multicolor fluorescence and X-ray computed tomography probe and application for non-invasive functional gastrointestinal tract imaging. Chem Mater 24:3772–3779. https://doi.org/10.1021/cm3023677
23. Zhuang P, Xiang K, Meng X, Wang G, Li Z, Lu Y, Kan D, Zhang X, Sun SK (2021) Gram-scale synthesis of a neodymium chelate as a spectral CT and second near-infrared window imaging agent for visualizing the gastrointestinal tract in vivo. J Mater Chem B 9:2285–2294. https://doi.org/10.1039/D0TB02276D
24. Shuai T, Zhou Y, Shao G, Yang R, Wang L, Wang J, Sun J, Ren L, Wang J, Liao Y et al (2020) Bimodal molecule as NIR-CT contrast agent for hepatoma specific imaging. Anal Chem 92:1138–1146. https://doi.org/10.1021/acs.analchem.9b04212
25. Men XJ, Chen HB, Sun C, Liu YB, Wang RB, Zhang XJ, Wu CF, Yuan Z (2020) Thermosensitive polymer dot nanocomposites for trimodal computed tomography/photoacoustic/fluorescence imaging-guided synergistic chemo-photothermal therapy. ACS Appl Mater Interfaces 12:51174–51184. https://doi.org/10.1021/acsami.0c13252
26. Oh MH, Lee N, Kim H, Park SP, Piao Y, Lee J, Jun SW, Moon WK, Choi SH, Hyeon T (2011) Large-scale synthesis of bioinert tantalum oxide nanoparticles for X-ray computed tomography imaging and bimodal image-guided sentinel lymph node mapping. J Am Chem Soc 133:5508–5515. https://doi.org/10.1021/ja200120k
27. Yu MX, Ma JJ, Wang JM, Cai WG, Zhang Z, Huang B, Sun MY, Cheng QY, Zhang ZL, Pang DW et al (2020) Ag_2Te quantum dots as contrast agents for near-infrared fluorescence and computed tomography imaging. ACS Appl Nano Mater 3:6071–6077. https://doi.org/10.1021/acsanm.0c01274
28. Qin MY, Yang XQ, Wang K, Zhang XS, Song JT, Yao MH, Yan DM, Liu B, Zhao YD (2015) In vivo cancer targeting and fluorescence-Ct dual-mode imaging with nanoprobes based on silver sulfide quantum dots and iodinated oil. Nanoscale 7:19484–19492. https://doi.org/10.1039/C5NR05620A
29. Chen LL, Zhao L, Wang ZG, Liu SL, Pang DW (2022) Near-infrared-II quantum dots for in vivo imaging and cancer therapy. Small 18:2104567. https://doi.org/10.1002/smll.202104567
30. McHugh KJ, Jing LH, Behrens AM, Jayawardena S, Tang W, Gao MY, Langer R, Jaklenec A (2018) Biocompatible semiconductor quantum dots as cancer imaging agents. Adv Mater 30:1706356. https://doi.org/10.1002/adma.201706356
31. Nieves LM, Mossburg K, Hsu JC, Maidment ADA, Cormode DP (2021) Silver chalcogenide nanoparticles: a review of their biomedical applications. Nanoscale 13:19306–19323. https://doi.org/10.1039/D0NR03872E
32. Hong G, Lee JC, Robinson JT, Raaz U, Xie L, Huang NF, Cooke JP, Dai HJ (2012) Multifunctional in vivo vascular imaging using near-infrared II fluorescence. Nat Med 18:1841–1846. https://doi.org/10.1038/nm.2995
33. He XW, Ma N (2014) An overview of recent advances in quantum dots for biomedical applications. Colloids Surf B Biointerfaces 124:118–131. https://doi.org/10.1016/j.colsurfb.2014.06.002
34. Du YP, Xu B, Fu T, Cai M, Li F, Zhang Y, Wang QB (2010) Near-infrared photoluminescent Ag_2S quantum dots from a single source precursor. J Am Chem Soc 132:1470–1471. https://doi.org/10.1021/ja909490r
35. Zhang Y, Hong GS, Zhang YJ, Chen GC, Li F, Dai HJ, Wang QB (2012) Ag_2S quantum dot: a bright and biocompatible fluorescent nanoprobe in the second near-infrared window. ACS Nano 6:3695–3702. https://doi.org/10.1021/nn301218z
36. Hong GS, Robinson JT, Zhang YJ, Diao S, Antaris AL, Wang QB, Dai HJ (2012) In vivo fluorescence imaging with Ag_2S quantum dots in the second near-infrared region. Angew Chem 51:9956–9959. https://doi.org/10.1002/anie.201206059
37. Li CY, Zhang YJ, Chen GC, Hu F, Zhao K, Wang QB (2017) Engineered multifunctional nanomedicine for simultaneous stereotactic chemotherapy and inhibited osteolysis in an orthotopic model of bone metastasis. Adv Mater 13:1605754. https://doi.org/10.1002/adma.201605754
38. Nieves LM, Hsu JC, Lau KC, Maidment ADA, Cormode DP (2021) Silver telluride nanoparticles as biocompatible and enhanced contrast agents for X-ray imaging: an in vivo breast cancer screening study. Nanoscale 13:163–174. https://doi.org/10.1039/D0NR05489E
39. deKrafft KE, Xie Z, Cao G, Tran S, Ma L, Zhou OZ, Lin W (2009) Iodinated nanoscale coordination polymers as potential contrast agents for computed tomography. Angew Chem 121:10085–10088. https://doi.org/10.1002/anie.200904958

40. Haller C, Hizoh I (2004) The cytotoxicity of iodinated radiocontrast agents on renal cells in vitro. Invest Radiol 39:149–154. https://doi.org/10.1097/01.rli.0000113776.87762.49
41. Wang Y, Wu YY, Liu YJ, Shen J, Lv L, Li LB, Yang LC, Zeng JF, Wang YY, Zhang LW, Li Z, Gao MY, Chai ZF (2016) BSA-mediated synthesis of bismuth sulfide nanotheranostic agents for tumor multimodal imaging and thermoradiotherapy. Adv Funct Mater 26:5335–5344. https://doi.org/10.1002/adfm.2016013
42. Hu DH, Sheng ZH, Zhang PF, Yang DZ, Liu SH, Gong P, Gao DY, Fang ST, Ma YF, Cai LT (2013) Hybrid gold-gadolinium nanoclusters for tumor-targeted NIRF/CT/MRI triple-modal imaging in vivo. Nanoscale 5:1624–1628. https://doi.org/10.1039/C2NR33543C
43. Wang JM, Zhu CN, Zhu DL, Tian ZQ, Lin Y, Pang DW (2015) Synthesis of photoluminescence Ag_2Te quantum dots in second near-infrared window. Chem J Chin U 36:1264–1268.
44. Ma JJ, Yu MX, Zhang Z, Cai WG, Zhang ZL, Zhu HL, Cheng QY, Tian ZQ, Pang DW (2018) Gd-DTPA coupled Ag_2Se quantum dots for dual-modality magnetic resonance imaging and fluorescence imaging in the second near-infrared window. Nanoscale 10:10699–10704. https://doi.org/10.1039/C8NR02017E

Chapter 13

Quantum Dot-Based Approaches for Labeling Epithelial Damage in Living and Fixed Zebrafish Embryos

Valeria Sigot

Abstract

Quantum dots (QDs) have been widely used in vivo for long-term monitoring in mammals and, more recently, for labeling zebrafish at early developmental stages. Here, it is described two protocols using carboxyl-coated QDs, which exhibit different labeling patterns depending on whether incubation is performed in vivo or after fixation of zebrafish embryos. Both protocols are presented in the context of an epithelial disruption model of the epidermis using EGTA (ethylene glycol-bis(β-aminoethyl ether)-*N,N,N′,N′*-tetraacetic acid), highlighting differences between damaged and undamaged epithelial tissues based on the distribution and labeling patterns of QDs.

Keywords Carboxyl-coated QDs, EGTA, Epithelial disruption, Zebrafish embryos, Apical microridges, Fluorescence microscopy

1 Introduction

The zebrafish (*Danio rerio*) is a widely used model for studying wound closure and epithelial regeneration because its embryos are optically transparent, develop externally, and can be monitored in vivo at high resolution [1–3]. Epithelial proteins in zebrafish share high functional similarity with those of mammals, further supporting its translational relevance [4, 5].

Fluorescent transgenic strains allow for specific monitoring of epithelial cells; however, their acquisition and maintenance can be limiting for small laboratories and aquaria. Therefore, QDs owing to their high brightness, resistance to photobleaching, and stability in aqueous environments, provide an attractive alternative for in vivo labeling, as well as for fixed embryos. In zebrafish, reports of vital labeling with QDs—either by exogenous incubation in the embryo medium or microinjection at one cell stage—remain limited, largely due to concerns about toxicity [6], and have mainly focused on the vascular and neuronal systems [6–8].

Goreti Pereira et al. (eds.), *Quantum Dots*, Methods in Molecular Biology, vol. 3025,
https://doi.org/10.1007/978-1-0716-5210-7_13,

In zebrafish embryos, the developing epidermis serves as the sole barrier to the environment. Generating precise and reproducible epithelial wounds at this stage remains a challenge and most studies employ mechanical or thermal ablation [9]. An alternative to these methods is chemical ablation using egtazic acid (EGTA) (ethylene glycol-bis(β-aminoethyl ether)-*N,N,N′,N′*-tetraacetic acid), a Ca^{2+} chelating agent that interferes with calcium dependent junctional proteins. Previously, we tested 3.4 nm carboxyl-coated CdTe QDs on intact and damaged epidermis after exposure to EGTA, and observed two distinct labeling patterns depending on whether embryos were incubated in vivo or postfixation after removing EGTA [10, 11]. In vivo incubation revealed that EGTA alters the paracellular transport pathway, affecting the passage of macromolecules through the epithelium. Under these conditions, high-intensity QDs fluorescence was observed, but cellular boundaries could not be resolved. In contrast, after fixation of damaged embryos, QDs revealed a distinct pattern of apical microridges in undamaged epidermis, while detached rounded cells were detected in damaged areas. This indicated that carboxyl-coated QDs can be used as sensitive markers of epithelial barrier function in vivo, as well as the integrity of the epidermis in fixed samples. Similar results were obtained with commercially available carboxyl-coated QDs (15–20 nm).

This chapter describes two complementary protocols for labeling zebrafish embryos with carboxyl-coated QDs in the context of an epithelial disruption model. The procedures could be applied either in vivo to monitor dynamic cellular responses, or in fixed embryos to achieve higher-contrast imaging, and may be carried out separately or in parallel depending on the experimental objective.

2 Materials

Prepare all solutions using ultrapure water and analytical grade reagents. Prepare all reagents at room temperature (approximately 25°C) and store at 4°C (unless indicated otherwise).

2.1 Zebrafish Embryos

1. Collect wild-type zebrafish (*Danio rerio*) embryos of T/AB genetic background at one-cell stage.
2. Rinse embryo with Ringer's solution 0.1× prewarmed at 29°C and transfer to plastic Petri dishes (10 embryos/30 mL Ringer's solution 0.1×).
3. Incubate overnight at 29°C until proper developmental stage of 24 h postfertilization (hpf) [12].

2.2 Stock and Working Samples

1. Stock carboxyl-coated QDs suspension: 8 μmol/L, mean diameter between 3 and 20 nm (*see* **Note 1**).

2. Stock phosphate-buffered saline (PBS) 1×, sterile, pH 7.8. Working solution: dilute 1× PBS to 0.1× with sterile distilled water prior to experiments.
3. Stock Ringer's solution 1× (111 mmol/L NaCl, 5 mmol/L KCl, 1 mmol/L $CaCl_2$, 0.6 mmol/L $MgSO_4$; sterile, pH 7.8). Working solution: dilute 1× Ringer to 0.1× with sterile distilled water prior to experiments.
4. Carboxyl-QDs working suspension for in vivo labeling: dilute carboxyl-coated QDs stock to 100–500 nmol/L in 0.1× PBS (*see* **Note 2**).
5. Carboxyl-coated QDs working suspension for fixed embryos: dilute carboxyl-coated QDs stock to 100–500 nmol/L in 1× PBS.
6. Stock solution of 500 mmol/L EGTA (prepared from acid form MW: 380 g/mol) in water, pH 8. Weigh 9.51 g EGTA powder. Add to a 100 mL beaker with ~35 mL distilled water. Adjust final pH to 8.0, transfer to a graduated cylinder and bring final volume to 50.0 mL with distilled water, filter sterilize, and aliquot 1.5 mL microtubes, store at –20 °C. Working solution: dilute stock solution 1:20 with 0.1× PBS to 25 mmol/L EGTA (*see* **Note 3**).
7. 1% w/v Low-melting agarose: prepare in distilled water by dissolving 0.2 g of low melting agarose in 20 mL of distilled water in a glass bottle. Heat in a microwave until fully melted. Store at room temperature and remelt in the microwave as needed.
8. 0.02% (w/v) tricaine in 1× Ringer's solution pH 7.8: prepare 20 mL. Store 1 mL aliquots at –20°C and thaw before use.
9. Sterile Glycerol–PBS Solution: Prepare 10 mL of 50% glycerol in 1× PBS. Sterilize by filtration (0.22 μm filter) and store at 4°C.
10. Carnoy's solution (ethanol:chloroform:acetic acid, 6:3:1). Prepare Carnoy's fixative under a fume hood. For 100 mL final volume, measure 60 mL absolute ethanol into a clean glass beaker. Add 30 mL chloroform slowly while mixing. Finally, add 10 mL glacial acetic acid while stirring. Mix thoroughly. Use immediately or store in a tightly sealed amber glass bottle at 4°C (*see* **Note 4**).

2.3 Equipment

1. Incubator set to 29°C for fish embryos.
2. Plastic Petri dishes, 90 mm, for embryo collection and incubation at 29°C.
3. Stereomicroscope for embryo handling and screening of epithelial damage.

4. Micropipettes (P10–P1000) and sterile tips.
5. 1.5 mL polypropylene microcentrifuge tubes.
6. 12-well plates for EGTA exposure.
7. Pasteur pipettes for transfer of embryos from Petri dishes to wells or to microcentrifuge tubes.
8. Glass-bottom 35 mm dishes or chambered coverslips for microscopy.
9. Fine forceps or tungsten needles to orient embryos in glass bottom chambers after covering with low melting agarose (*see* **Note 5**).
10. Wide field fluorescence sectioning microscope or confocal microscope equipped with, for example, UPLFLN 20×, NA 0.75; UPLFLN 40× NA 1.3 oil (for fixed embryos) or NA 1.2 water.
11. Vortex.
12. Bench microcentrifuge.
13. Fume hood for fixative preparation.

3 Methods

3.1 Zebrafish Embryo Handling

1. For the experimental procedures use embryos at pharyngula stage or in the range of 27–33 hpf [12].
2. At the end of experiments, euthanize embryos by immersion in 0.02% tricaine, for 20 min at 4°C (*see* **Note 6**).
3. 0.1× PBS and 0.1× Ringer's solution have conductivity in the range for zebrafish embryo rearing (400–600 total dissolved salts). Warm PBS 0.1× and Ringer 0.1× to 29°C in the incubator before use.

An overview of the experimental workflow is presented in Fig. 1.

3.2 Induction of Epithelial Damage by EGTA Treatment

1. At 24 hpf, remove chorion using fine forceps or pronase, if necessary, and replace 0.1× Ringer's solution with previously warmed 0.1× PBS. Maintain embryos at 29°C until 27 hpf.
2. To a 12 well-plate transfer two embryos per well to 8 wells (two rows of 4 wells for EGTA-treated) and transfer 2 embryo per well to 4 wells (one row for controls).
3. EGTA treatment: Replace the 0.1× PBS with a working solution of 25 mM EGTA in 0.1× PBS and incubate for at least 15 min at 29°C to induce epithelial disruption. This step requires monitoring under the stereoscope (*see* **Note 7**).

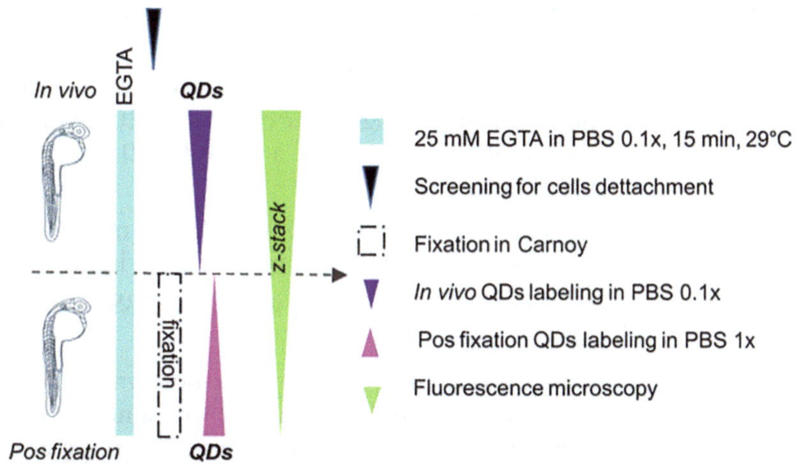

Fig. 1 Schematic overview of the experimental workflow for QD labeling in vivo and postfixation. Zebrafish embryos are subjected to EGTA treatment in 0.1× PBS to induce epithelial injury, followed by two labeling approaches: [1] in vivo labeling, where live embryos are incubated with QDs in 0.1× PBS immediately after EGTA exposure to visualize epithelial cell detachment, and (i) postfixation labeling, where embryos are fixed overnight in Carnoy's solution and subsequently exposed to QDs in 1× PBS for higher-resolution imaging. After labeling, live embryos are mounted in glass-bottom dishes with low-melting agarose and 0.02% tricaine, whereas fixed embryos are mounted in glycerol (50%) in 1x PBS

4. Leave control embryos in 0.1× PBS, 15 min at 29°C.
5. Rinse embryos three times with 0.1× PBS to remove residual EGTA.
6. Transfer 8 treated embryos (2×4, first row) and four control untreated embryos to individual sterile 1.5 mL polypropylene micro tubes in 0.1× PBS.
7. Follow immediately with the *in vivo* QD labeling approach (Sect. 3.3).
8. Transfer the rest of the EGTA-treated embryos (2×4 embryos, second row) and 4 control embryos from the well plate to individual sterile 1.5 mL polypropylene microtubes (after Step 5).
9. Carefully remove with a 20 μL pipette the excess of the 0.1× PBS and replace with 500 μL Carnoy's solution under the hood.
10. Discard the Carnoy and add a fresh aliquot of 500 μL; repeat this wash until the solution remains transparent.
11. Incubate embryos in fresh Carnoy´s fixative overnight at 4°C.
12. Store embryos until further use for QD labeling of fixed embryos (Sect. 3.4).

3.3 QD Labeling of Live Embryos

1. Prepare 1 mL of QDs working suspension of 100–500 nmol/L in 0.1× PBS.
2. Remove with a 20 μL pipette the excess of 0.1× PBS from the tubes containing either control or EGTA-treated embryos (four embryos per tube) and add 100 μL of QDs working suspension.
3. Incubate the tubes horizontally for 15 min at 29°C to avoid embryo superposition and protect from light.
4. Rinse embryos three times with 0.1× PBS to remove unbound QDs.
5. Transfer four embryos to a glass-bottom dish with 200 μL of 0.1× Ringer´s solution with 0.02% tricaine and mount them by adding low melting agarose with a 2–10 μL pipette, this should be melted with loosen cap in a microwave oven at low power. Before covering the embryos and adding tricaine, verify that the agarose temperature is skin-tolerable.
6. Orient embryos laterally with a tungsten needle, remove excess agarose, allow it to solidify at 25°C, and then cover with additional 0.1× Ringer's solution containing 0.02% tricaine for image acquisition (see Sect. 3.5).

3.4 QD Labeling of Fixed Embryos

1. Prepare 1 mL of QDs working suspension of 100–500 nmol/L in 1× PBS.
2. Place tubes containing fixed embryos, (controls and EGTA-treated) under a chemical hood and carefully remove Carnoy's solution using a pipette.
3. Add 1 mL of 1× PBS to each tube (four embryos per tube), gently invert or tap the tube to wash; allow embryos to settle by gravity.
4. Remove PBS without disturbing embryos and repeat the wash 3–4 times, or until the supernatant appears clear (*see* **Note 8**).
5. Remove 1× PBS from the tubes and add 100 μL of carboxyl-coated QDs working suspension for 30 min at 25°C protected from light. Incubate the tubes horizontally to avoid embryo superposition and protect from light.
6. Rinse embryos three times in 1× PBS and mount in 50% glicerol-1× PBS on glass-bottom dishes for imaging. Protect from direct light exposure.

7. Leave the glass-bottom dishes with mounted embryos overnight at 4°C to improve tissue clarification before image acquisition.

3.5 Image Acquisition

3.5.1 For Live Embryos Mounted in 1% Low-Melting Agarose

1. Acquire images (z-stacks) using a wide field fluorescence sectioning microscope.
2. Perform first bright-field imaging with 4× and 10× objectives to evaluate global embryo integrity and EGTA damaged regions, mainly in the tail tip (Fig. 2).
3. For orange-red emitting QDs detect fluorescence with excitation filter ~540–550 nm and emission filter ~570–620 nm (*see* **Note 9**).
4. Use 20× objectives NA 0.75 to acquire a z-stack covering the epidermal bilayer (~6.6 μm) with at least 10 optical sections at 1 μm intervals. This will show QDs penetration only in the damaged region upon epithelial disruption with EGTA (Fig. 2).

3.5.2 For Fixed Embryo Mounted in Glycerol 50%-1×PBS

1. *Perform bright-field imaging with 4×* and 10× objectives to evaluate embryo integrity and EGTA damaged regions.
2. *Detect red QD fluorescence as in Step 3. For high-resolution imaging, use a 40×* oil immersion objective (NA 1.3). This allows detailed visualization of apical microridge patterns in normal epidermis and of cell rounding and microridge loss in damaged regions of the external epidermal layer (Fig. 3).

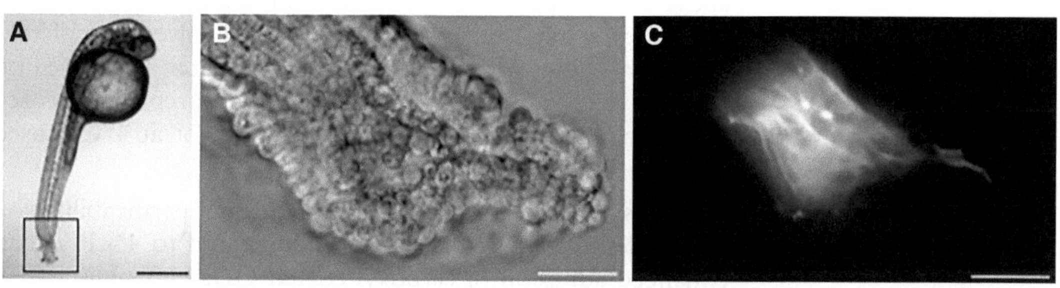

Fig. 2 In vivo QDs labeling of epithelial damage in the early finfold of zebrafish embryo. (**A**) Bright-field image of a zebrafish embryo at stage 27 hpf after EGTA treatment (boxed region indicates the damaged region at the embryo tail tip). (**B**) Higher magnification of the damaged fin fold with round and detached epidermal cells. (**C**) In vivo QDs labeling highlights disrupted tissue integrity without distinction of individual cells. Scale bars: (**A**) 200 μm (**B–C**) 50 μm

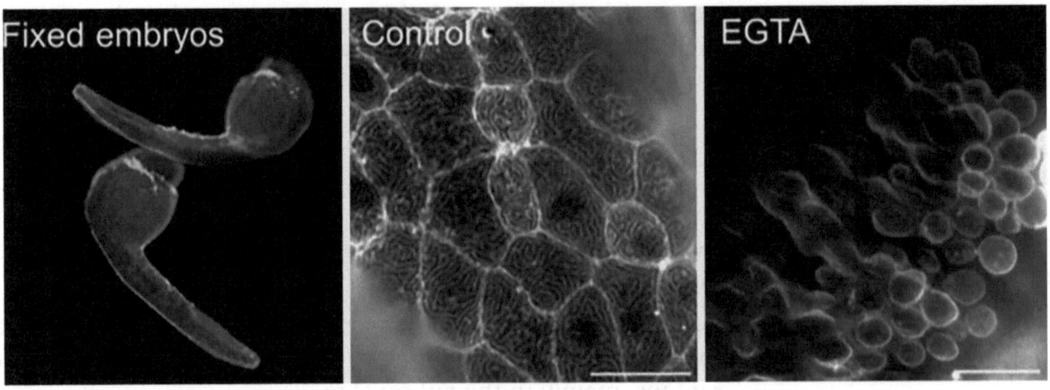

Fig. 3 Postfixation-QDs labeling of epithelial damage in the early finfold of zebrafish embryo. (**A**) Zebrafish embryos surface labeled with QDs. (**B**) Normal epidermis in control and fixed embryo, where QDs staining highlights at the subcellular level the apical microridge pattern on the epidermis. (**C**) In EGTA-treated and fixed embryos, QDs labeling reveals the loss of microridges upon cell roundness and detachment. Scale bars: (**A**) 200 μm (**B–C**) 20 μm

4 Notes

1. QDs are stable in 1× PBS and 0.1× PBS pH 7.8–8.5. Most commercial QDs contains cadmium and selenium in an inorganic crystalline form. Dispose these classes of material in compliance with all applicable local, state, and federal regulations.
2. Vortex the stock carboxyl-coated QDs suspension and adjust QD concentration to 100–500 nmol/L depending on labeling brightness and background. Excessive QD concentration may result in nonspecific aggregation and toxicity.
3. The acid form of EGTA will not dissolve until the pH is raised above 7.5. Slowly add NaOH solution (e.g., 5 mol/L NaOH, dropwise) while stirring until EGTA dissolves completely. (Optional) Filter with a 0.22-μm sterilized filter. Dispense into aliquots (e.g., 1 mL or 5 mL tubes) to avoid repeated freeze–thaw cycles. Store at −20°C for long term, or at 4°C if used within 1–2 weeks.
4. Carnoy's fixation does not require further permeabilization step and shows lower background compared to 4% PFA and enhanced adhesion of carboxyl-coated QDs [10]. Dispose of Carnoy waste according to institutional safety regulations.
5. Embryos mounted in agarose should be oriented laterally to facilitate observation of the epidermis in the trunk and tail region.

6. All experiments must comply with institutional animal care protocols and adhere to national and international ethical guidelines.
7. Optimization of EGTA treatment is critical: exposure >30 min at 25 mmol/L EGTA can cause irreversible tissue damage and embryo lethality. Assess tissue integrity and survival under a stereomicroscope. The endpoint of EGTA treatment is characterized by clusters of round and detached epithelial cells, which first appear at the tail tip within the early finfold [11]. Sublethal toxicity includes edema, deformation, or yolk sac abnormalities. Only experiments in which control embryo survival is ≥ 85% should be considered valid.
8. After removing Carnoy's solution embryos will appear white and opaque, which indicates proper fixation.
9. Red-emitting QDs are desirable, as they are easily distinguishable from zebrafish autofluorescence. Minimize light exposure during excitation using the lowest potency of the laser that gives a suitable signal to noise ratio to decrease phototoxicity for acquisition in vivo, especially during extended time-lapse imaging (e.g., more than 1 h).

Acknowledgments

This work was supported by Agencia Nacional de Promoción Cientıfica y Técnica Grant PICT 020-2594 and Grant PID-UNER 6191 from National University of Entre Rıos and National Scientific and Technical Research Council (CONICET-Argentina).

References

1. Godinho L (2011) Live imaging of zebrafish development. Cold Spring Harb Protoc 2011(7):770-7. https://doi.org/10.1101/pdb.top119. PMID: 21724823
2. Howe K, Clark MD, Torroja CF, Torrance J, Berthelot C, Muffato M et al (2013) The zebrafish reference genome sequence and its relationship to the human genome. Nature 496:498–503. https://doi.org/10.1038/nature12111
3. Castranova D, Samasa B, Venero Galanternik M, Gore AV, Goldstein AE, Park JS et al (2022) Long-term imaging of living adult zebrafish. Development 149:dev199667. https://doi.org/10.1242/dev.199667
4. Le Guellec D, Morvan-Dubois G, Sire J-Y (2003) Skin development in bony fish with particular emphasis on collagen deposition in the dermis of the zebrafish (*Danio rerio*). Int J Dev Biol 48:217–231
5. Rakers S, Gebert M, Uppalapati S, Meyer W, Maderson P, Sell AF et al (2010) 'Fish matters': the relevance of fish skin biology to investigative dermatology. Exp Dermatol 19:313–324. https://doi.org/10.1111/j.1600-0625.2009.01059.x
6. Rieger S, Kulkarni RP, Darcy D, Fraser SE, Köster RW (2005) Quantum dots are powerful multipurpose vital labeling agents in zebrafish embryos. Dev Dyn 234:670–681. https://doi.org/10.1002/dvdy.20524
7. King-Heiden TC, Wiecinski PN, Mangham AN, Metz KM, Nesbit D, Pedersen JA et al (2009) Quantum dot nanotoxicity assessment using the zebrafish embryo. Environ Sci Technol 43:1605–1611. https://doi.org/10.1021/es801925c

8. Son SW, Kim JH, Kim SH, Kim H, Chung A, Choo JB et al (2009) Intravital imaging in zebrafish using quantum dots. Skin Res Technol 15:157–160. https://doi.org/10.1111/j.1600-0846.2008.00341.x
9. Marques IJ, Lupi E, Mercader N (2019) Model systems for regeneration: zebrafish. Development 146:dev167692. https://doi.org/10.1242/dev.167692
10. Sigot V, Cabral Filho PE, Sampedro MF, Santos BS, Fontes A (2020) Anionic Quantum dots reveal actin-microridges in zebrafish epidermis. Methods Appl Fluoresc 8:035007. https://doi.org/10.1088/2050-6120/ab9124
11. Sampedro MF, Sian ME, Filho PEC, Fontes A, Sigot V (2025) EGTA-induced disruption of adherens junctions in zebrafish embryonic epidermis. FEBS Lett 599:3456–3469. https://doi.org/10.1002/1873-3468.70142
12. Kimmel CB, Ballard WW, Kimmel SR, Ullmann B, Schilling TF (1995) Stages of embryonic development of the zebrafish. Dev Dyn 203:253–310. https://doi.org/10.1002/aja.1002030302

Part IV

Detection and Sensing

Check for updates

Chapter 14

Quantum Dots for Assessment of ROS Accumulation During Chemotherapy, Radiotherapy, and Chemoradiotherapy

Jackie Austin, Kimal Honour Djam, Bong Han Lee, Sindhuja Suresh and Andrew Edet Ekpenyong

Abstract

Quantum dots (QDs) are semiconductor nanoparticles ranging in size from 2 to 10 nm. QDs are increasingly being developed for biomedical imaging, targeted drug delivery, and green energy technology. Here, we describe the novel utilization of biocompatible (CdSe)/ZnS core/shell semiconductor nanoparticles for assessment of reactive oxygen species (ROS) in the context of chemotherapy and radiotherapy, both of which are important modalities in the treatment of cancer. Computational simulations done using COMSOL Multiphysics confirm some of the properties of the QDs used in the described experiments.

Keywords Reactive oxygen species, Quantum dot, Fluorescence, Cancer, Doxorubicin, Daunorubicin, Therapy, COMSOL multiphysics, Schrodinger equation

1 Introduction

Single quantum dots (QD) can be made from as few as 100 to as many as 100,000 atoms of semiconductor elements via colloidal synthesis, such that joining these dots into complex assemblies enables novel discoveries and applications [1]. QDs hold distinct properties and diverse applications due to their size-dependent quantum effects [2–4]. Although semiconductor theory describes the bulk material properties of QDs, quantum mechanics is needed to describe several optoelectronic properties of QDs. For instance, the time-independent Schrödinger equation can be solved to obtain energy eigenvalues for QDs using potential wells, such as infinite square wells. Such energy eigenvalues, along with energy band gap and Coulombic interactions, provide a theoretical framework that accounts for size of the QDs and their excitation and emission wavelengths [5, 6].

QDs continue to have myriad applications in systems as varied as photovoltaic cells and biological cells, quantum information

Goreti Pereira et al. (eds.), *Quantum Dots*, Methods in Molecular Biology, vol. 3025,
https://doi.org/10.1007/978-1-0716-5210-7_14,

processing, and targeted drug delivery. For biomedical applications, there is increasing attempt to develop and use QD fluorescent probes in cellular environments and so much effort has been devoted to assessing the toxicity of QDs on biological cells [7–10]. Interestingly, biological cells themselves produce molecules whose quantities and availabilities following interactions with the cellular environment control cellular functions and determine cellular fates. Reactive oxygen species (ROS) are eminent among such molecules.

ROS are chemically unstable molecules that contain oxygen but are more reactive than molecular oxygen [11]. ROS were first described only 60 years ago [12] and their roles in cell physiology and pathology are still being clarified [13]. They were first seen as toxic molecules produced by phagocytes to destroy bacteria and other pathogens [14]. However, it is now known that all cell types produce ROS [11, 13]. Moreover, certain levels of ROS are beneficial to the cell, while certain levels are harmful [11]. Of course, if the mechanisms of ROS production at the cellular level were well understood, the harmful levels could be altered as a therapeutic measure, but such mechanisms are not well understood [15]. Hence, old probes continue to be refined, and new probes are being developed for the detection and quantification of ROS from living organisms [16] so as to ultimately unravel the mechanisms of ROS production and also control cell fate.

Probes for the detection of ROS can be categorized into calorimetric, electrochemical, and fluorescent sensors [17]. QD-based probes for ROS detection belong to the category of fluorescent sensors. The development of QD fluorescent probes for cellular ROS measurements is a rapidly burgeoning enterprise [9, 17–20]. Understandably, these new QD fluorescent probes are meant to overcome challenges posited by ROS such as their very short lifetimes (around 10^{-6}s), variety and competition between ROS of cellular origin and those from bioactive molecules such as drugs [17, 21, 22].

Being able to quantitatively differentiate ROS levels due to bioactive molecules from drugs and ROS levels of cellular origin is important for nanoparticle mediated drug delivery (NPDD), especially cancer drug delivery [23, 24], for chemotherapy and for radiotherapy, including nanoparticle-mediated radiotherapy (NPRT). For example, QDs have been conjugated with the chemotherapeutic drug, doxorubicin (Dox), for targeted delivery against lung cancer [25]. QDs themselves are known to generate or modulate ROS production in cells making them candidate radiosensitizers against cancer cells that are radioresistant [12, 26].

We have joined the quest to develop QD based ROS sensors, assess QD toxicity on cells, use them for NPDD, and as radiosensitizers by developing a fluorescence spectroscopy assay (Fig. 1), which shows that ROS from chemotherapeutic drugs, Dox, and daunorubicin (Dauno) in cell culture medium without cells, significantly

($p < 0.0001$) modulate the fluorescence intensity of CdSe/ZnS QDs [6, 22] (Fig. 2).

Interestingly, this modulation is upheld, but at a different level, when cancer cells lines, HL60 (myeloid), and K562 (erythroid) are treated with the Dox and Dauno (Fig. 3), a result that engenders the simultaneous QD labeling and imaging of erythroid and myeloid cells while using the QDs to enhance drug uptake during chemotherapy.

Furthermore, and in agreement with our previously published work [22], our QD assay detects significantly ($p < 0.0001$) enhanced ROS production in irradiated radiosensitive cells (HL60 and K562

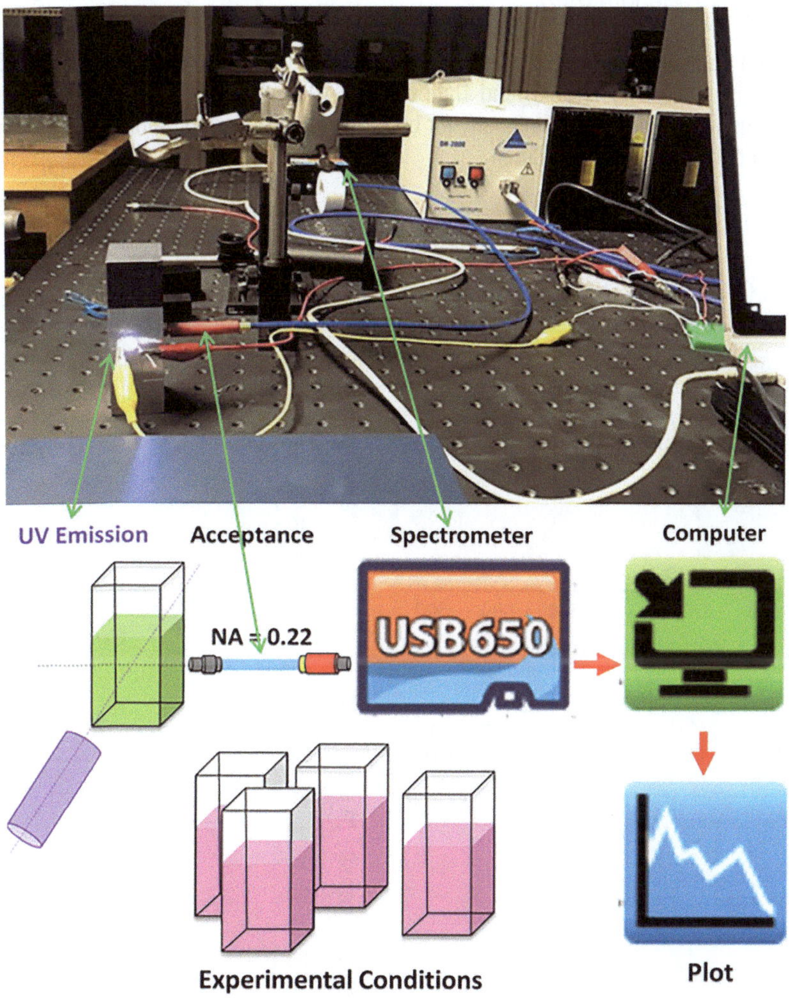

Fig. 1 Spectroscopy equipment and setup. The colloidal solution of CdSe/ZnS QD inside the cuvette is illuminated with UV from the LED at a temperature of 22°C. Fluorescence from the QD is picked up by the optical fiber, gets processed in a spectrometer, and analyzed in the computer. (The setup shown here is upgraded and more robust than the one illustrated in [22], thanks to the introduction of light shielding cuvette holder, and optional multiple light sources)

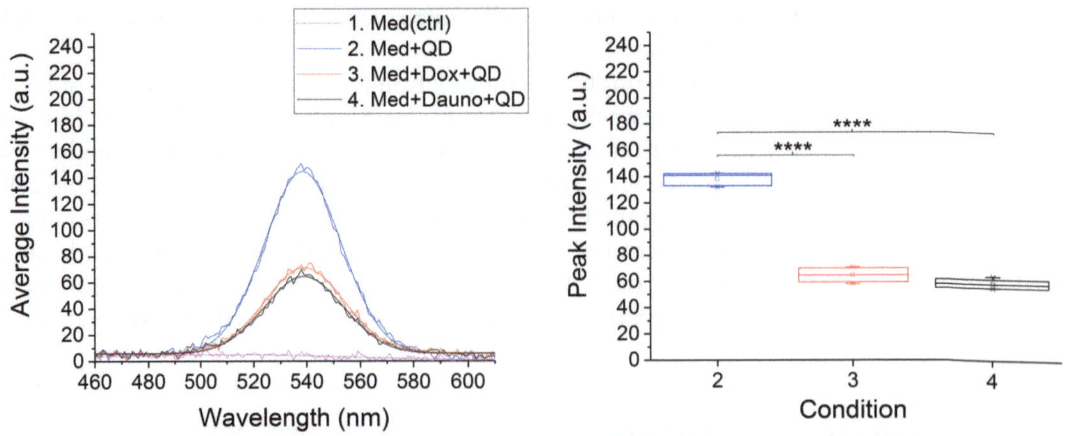

Fig. 2 Spectra and statistical comparison for hydrophilic PEGylated CdSe/ZnS QD with medium (Med) and chemotherapy drugs, Doxorubicin (Dox), and Daunorubicin (Dauno). There is statistically significant ($p < 0.0001$) reduction in peak fluorescence intensity for both Dox and Dauno–containing media (reproduced from [22] with permission from John Wiley & Sons)

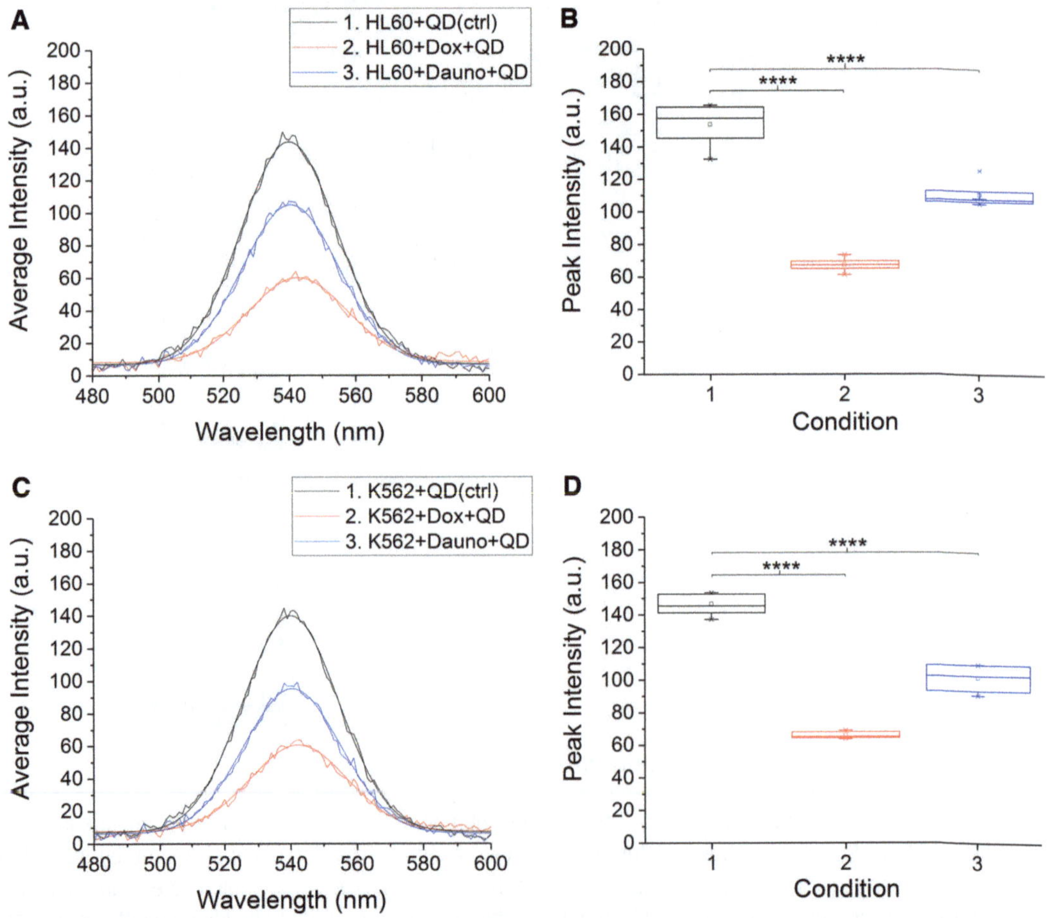

Fig. 3 Spectra and statistical comparison for cells treated with chemotherapy drugs. HL60 cells (**A, B**) and K562 cells (**C, D**) both show significant ($p < 0.0001$) intensity modulation of QDs for both Dox and Dauno conditions (reproduced from [22] with permission from John Wiley & Sons)

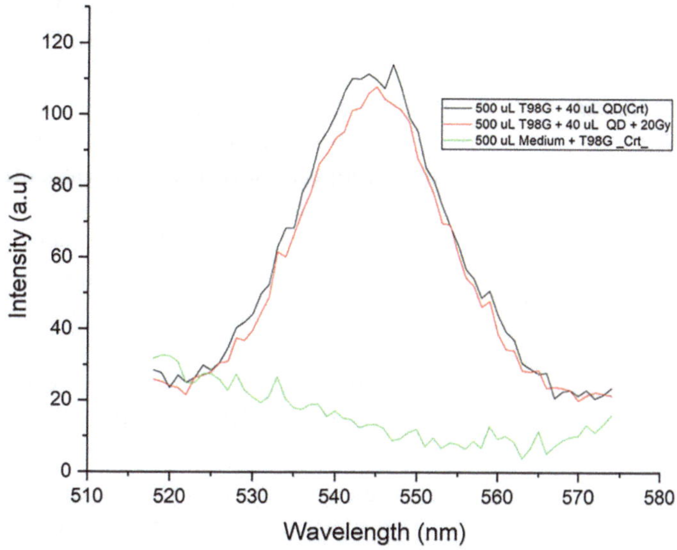

Fig. 4 Spectra for T98G cancer cells treated with 20-Gy of X-rays. The reduction in fluorescence intensity for these radioresistant cancer cells is much less than the case for radiosensitive cancer cells (HL60 and K562, see [22])

blood cancer cells) and a markedly lower level of such enhanced ROS production in irradiated radioresistant cells, T98G brain cancer cells (Fig. 4). This opens up a vista that enables using QDs for simultaneous assessment of ROS and radiosensitization for improvement of outcomes of radiotherapy.

2 Materials

Carryout all procedures using standard cell culture facilities such as biosafety cabinet class II. Prepare and store all reagents under sterile conditions to ensure purity and reproducibility. For simulations, the COMSOL Multiphysics application package and a computer are needed.

2.1 Spectroscopic and Radiotherapeutic Materials and Equipment

1. Our setup for fluorescence intensity measurements had a light-emitting diode (LED) (3.3 V, 0.2 A) in the ultraviolet (UV) region, an optical fiber cable (UV/VIS fiber type) with a numerical aperture of 0.22, providing an acceptance angle of 12.7° (*see* **Note 1**). The light is transformed into a signal in the spectrometer (USB 650, Ocean Optics). The signal is processed and analyzed in the computer using the spectrometer software. Figure 1 shows the spectroscopic experimental setup.
2. X-ray irradiator: Faxitron CellRad: a compact, safe cabinet X-ray unit used to apply clinically relevant doses of X-ray to

cells, where this induces ROS production in cells. This is in vitro radiotherapy (*see* **Note 2**).

2.2 Chemical Materials

1. Hydrophobic CdSe-ZnS QDs (emission wavelength, $\lambda_{em} = 540\,nm$) in powdered form: disperse the QDs in toluene, a nonpolar solvent, to a concentration of about 0.1 μmol/L, and use for control experiment to verify that the spectroscopic setup reads expected fluorescence.
2. Hydrophilic CdSe-ZnS QDs ($\lambda_{em} = 540\,nm$): prepare a stock suspension with the concentration of 0.16 μmol/L in deionized water, for the experiment with H_2O_2 to quantitate the setup (Sect. 3.1). For other experiments (Sects. 3.2–3.5, and 3.7), dilute directly with cell medium to obtain desired concentrations. Higher final concentrations are possible.
3. Preferentially use QDs coated with polyethylene glycol (PEG), an inert polymer that does not interfere with the optical properties of the QD. Store at 4°C. The PEGylation is not necessary for the current application but could be useful for clinical translation.
4. Hydrogen Peroxide H_2O_2: at 50% concentration from the manufacturer. Store at 4°C.
5. Dox stock solution with a concentration of 1 mmol/L in deionized water.
6. Dauno stock solution with a concentration of 0.2 mmol/L in deionized water.
7. Superoxide dismutase (SOD): from human erythrocytes (*see* **Note 3**). Store at 4°C.
8. Catalase (Cat): from bovine liver (*see* **Note 4**). Store at 4°C.
9. Chemotherapeutic Drugs: Dox. These common chemotherapeutic drugs induce controlled changes in cellular activities including enhanced ROS production and their overall effects on cells have been characterized by others [24, 27] and by us [28]. Protect from light by wrapping aliquots in aluminum foil. Store at −20°C.

2.3 Biological Materials

1. HL60 cells: human peripheral blood-derived acute promyelocytic leukemia cancer cells. Grow cells in suspension using standard methods and media, namely, RPMI 1640, supplemented with 10% fetal bovine serum (FBS) and 1% penicillin/streptomycin, inside incubator maintained at 95% air; 5% CO_2 and a temperature of 37°C (*see* **Note 5**).
2. K562 cells: human bone marrow derived chronic myelogenous leukemia, grow using the same media, and procedures as the HL60 cells (*see* **Note 5**).

3. T98G cells: human brain derived glioblastoma multiform cells with fibroblastic morphology. Grow them as prescribed using ATCC-formulated Eagle's Minimum Essential Medium and 10% FBS (*see* **Note 6**).

3 Methods

3.1 Mixing Colloidal Suspension of Hydrophilic CdSe-ZnS QDs with H_2O_2 to Quantitate the Setup

1. For 0% H_2O_2, take 520 μL of stock suspension of 0.16 μmol/L QD and add it into the cuvette.
2. For 0.1% H_2O_2, take 520 μL of stock suspension of QD, 1.04 μL of H_2O_2, and add them into the cuvette (*see* **Note** 7).
3. For 1% H_2O_2, take 518.96 μL of stock suspension of QD, 10.4 μL of H_2O_2, and add them into the cuvette.
4. For 5% H_2O_2, take 509.6 μL of stock suspension of QD 52 μL of H_2O_2, and add them into the cuvette.
5. For 10% H_2O_2, take 468 μL of stock suspension of QD 104 μL of H_2O_2, and add them into the cuvette.
6. Place the cuvette in the spectrometer setup and measure the fluorescence spectra. The result will be an inverse relationship between H_2O_2 concentration, and the QD fluorescence intensity, establishing optical quenching of QD as a function of ROS accumulation, thereby enabling ROS assessment.

3.2 Mixing Colloidal Suspension of Hydrophilic CdSe-ZnS QDs with Medium and Chemotherapeutic Drugs to Measure ROS Production Without Cells

1. Take out Dox and Dauno from the freezer (−20°C).
2. In two T25 flasks, add 5 mL of RPMI to each of them.
3. For a final concentration of 5 μmol/L of Dox, take 25 μL of Dox with an initial concentration of 1 mmol/L, and add it to one of the T25 flasks.
4. For a final concentration of 1 μmol/L of Dauno, take 25 μL of Dauno with an initial concentration of 0.2 mmol/L, and add it to the other T25 flask.
5. Take 500 μL from each of the T25 flasks and add them to the cuvettes.
6. Add 20 μL of hydrophilic 4 μmol/L (CdSe)/ZnS QDs to each of them, for a final concentration of 0.16 μmol/L and incubate for 1 h. Higher concentrations are possible.
7. Cover the cuvettes with Parafilm or similar plastic membrane.
8. Place the cuvette in the spectrometer setup and measure the fluorescence spectra.

3.3 Mixing Colloidal Suspension of Hydrophilic CdSe-ZnS QDs with Cells to Assess ROS Production by Cells

1. To add QDs onto cells, disinfect the hood before starting the experiment, and prepare a cuvette holder (*see* **Note 8** for the case of adherent cells such as T98G cells).
2. When using the mechanical pipette, use the 1000 μL sterilized pipette tip for the stock of cells.
3. Set the mechanical pipette to 500 μL, shake the T25 flask containing cells at 1 million cells/mL and press the plunger halfway of the mechanical pipette to extract 500 μL of cells.
4. Release the plunger to release the solution of cells onto the cuvettes.
5. Change the tip of the mechanical pipette with the 100 μL pipette tip.
6. Set the plunger and press to extract suspension of QDs, needed for desired final concentration.
7. Release the plunger to release the suspension of QD onto the cuvettes.
8. Press the plunger halfway 15 times to mix the combined suspension of cells and QD.
9. Cover the cuvettes with Parafilm or similar plastic membrane.
10. Place the cuvette in the spectrometer setup and measure the fluorescence spectra.

3.4 Mixing Colloidal Suspension of Hydrophilic CdSe-ZnS QDs with Cells Treated with Chemotherapeutic Drugs

1. Take out Dox and Dauno, as supplied, out of the freezer (−20°C).
2. In two T25 flasks, add 5 mL of cells to each of them.
3. For a final concentration of 5 μmol/L of Dox, take 25 μL of Dox with an initial concentration of 1 mmol/L, and add it to one of the T25 flasks.
4. For a final concentration of 1 μmol/L of Dauno, take 25 μL of Dauno with an initial concentration of 0.2 mmol/L, and add it to the other T25 flask.
5. Take 500 μL from each of the T25 flasks and add them to the cuvettes.
6. Add 20 μL of hydrophilic 4 μmol/L CdSe-ZnS QDs to each of them, for a final concentration of 0.16 μmol/L.
7. Press the plunger halfway 15 times to mix the combined suspension of cells, chemotherapeutic drugs, and QD.
8. Cover the cuvettes with Parafilm or similar plastic membrane.
9. Place the cuvette in the spectrometer setup and measure the fluorescence spectra.

3.5 Mixing Colloidal Suspension of Hydrophilic CdSe/ZnS QDs with Cells Post-Radiotherapy to Assess Radiation-Induced ROS Production

1. Warm up the X-ray irradiator cabinet.
2. Place cells that are in T25 flasks inside the X-ray irradiator cabinet. Note the level (distance from X-ray source and maintain that distance throughout all experiments).
3. Press automated dose control (ADC) mode to input the desired dose onto the cell culture.
4. Take 500 μL from each of the T25 flasks and add them to the cuvettes.
5. Add 20 μL of hydrophilic 4 μmol/L CdSe/ZnS QDs to each of them, for a final concentration of 0.16 μmol/L.
6. Place the cuvette in the spectrometer setup and measure the fluorescence spectra.

3.6 Using Hydrophobic CdSe-ZnS QDs to Test the Spectroscopic Setup

1. Use the hydrophobic CdSe-ZnS QDs suspension in toluene in the concentration of about 0.1 μmol/L to verify that the spectroscopic setup reads expected fluorescence.
2. Add about 500 μL of the suspension into a cuvette.
3. Place the cuvette in the spectrometer setup and measure the fluorescence spectra in order to read the peak wavelength.
4. Confirm that the peak wavelength is within the range given by the manufacturer of the QD and this should verify that the spectroscopic setup is working as expected.

3.7 Mixing Colloidal Suspension of Hydrophilic CdSe-ZnS QDs with SOD and CAT to Explore Mechanism of Intensity Modulation of QD Fluorescence

1. Carry out Steps 1–6 of Sect. 3.2.
2. Add SOD to a final concentration of 120 U/mL to both Dox and Dauno-containing media with QD, place in cuvettes (each containing about 500 μL of the suspension), and incubate for 2 h.
3. Place the cuvettes in the spectrometer setup and measure the fluorescence spectra. Compare the SOD containing QD suspension of media and Dox or media and Dauno with those without SOD. Since SOD breaks down the superoxide radical to molecular oxygen or hydrogen peroxide, there should be little or no difference when measured after 2 h of incubation with SOD. The modulation of fluorescence intensity should remain intact.
4. Incubate the mixture in Step 2 above for 24 h.
5. Following incubation with SOD for 24 h, add CAT to a final concentration of 1000 U/mL. Incubate for about 30 min and measure the fluorescence intensity.
6. Within 1 h of adding CAT, there should no longer be any fluorescence intensity modulation. The SOD should break down the superoxide radicals to molecular oxygen or hydrogen or hydrogen peroxide over the 24 h of incubation, while the CAT

should rapidly reduce any hydrogen peroxide still present to water and oxygen. The specificity of the SOD and CAT actions enable the molecular-level exploration of the ROS dynamics and some aspects of the mechanism of fluorescence intensity modulation.

3.8 Simulations Using COMSOL Multiphysics to Further Confirm Photophysical Properties of CdSe-ZnS QD

1. Set up the Model Environment in COMSOL, choosing the spatial dimension (3D for CdSe-ZnS), type of physics (Mathematics since Schrödinger Equation is to be solved as a PDE), and as a stationary study (time-independent).
2. Create geometry of the object. A spherical radius of 2.5 nm was used, with 1.5 nm outer shell thickness, and a 1 nm thick core.
3. Specify material properties. COMSOL uses the following PDE in coefficient form:

$$\nabla \cdot (-c\nabla u - \alpha u + \gamma) + \beta \cdot \nabla u + au = d_a \lambda u. \tag{1}$$

   Comparing Equation (Eq. 1) with the time-independent Schrödinger equation,

$$-\frac{\hbar^2}{2m}\nabla^2 \Psi + V\Psi = E\Psi, \tag{2}$$

   the terms α, β, and γ are 0 and so Equation (Eq. 1) reduces to

$$-c\nabla^2 + au = d_a \lambda u, \tag{3}$$

   and the nonzero coefficients are $c = \frac{\hbar^2}{2m}$, $a = V$ (the potential energy), $d_a = 1$, $\lambda = E$ (the total energy), and $u = \psi$ (the wave function). The values for the simulation parameters in the case of CdSe-ZnS used in the experiments and protocols reported here are found in Table 2.1 of [6].
4. Define physical boundary conditions. Dirichilet boundary conditions were placed at the boundaries of the surface of the QD. This confines the solutions of the PDE to be within the QD.
5. Create the mesh. Finite element method is used in COMSOL, thus, a mesh of tetrahedrons was added to the QD. The finer the mesh, the more exact the numerical solution.
6. Run the simulation. The simulation provides the energy eigenvalues of the CdSe-ZnS QD and probability density of the confined particle. The results are shown in Fig. 5 for electrons and in Fig. 6 for holes.
7. Postprocess the results. The resulting energy eigenvalues from the simulations are used to calculate the energy difference of

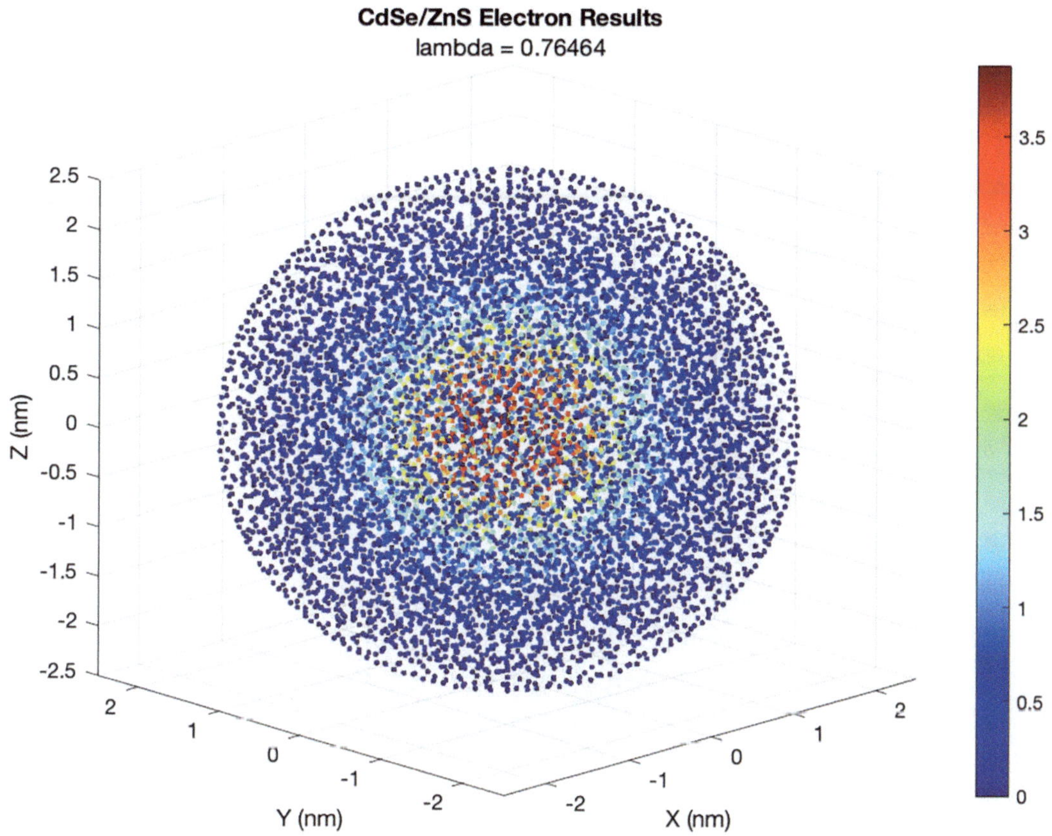

Fig. 5 Electron probability density for CdSe-ZnS. For this first energy eigenvalue, λ is 0.76464 eV and the radius is 2.5 nm

the QD (*see* **Note 9**). The lowest energy eigenvalues from the electron and hole were used for the energy difference ΔE calculation, as follows:

$$\Delta E = E_g + \frac{\hbar^2}{8m_e^* d^2} + \frac{\hbar^2}{8m_h^* d^2} - \frac{e^2}{8\pi\varepsilon_0 d_{eh}} \approx \frac{hc}{\lambda_{em}}, \tag{4}$$

where E_g is the energy band gap of the confined region (which for CdSe-ZnS is the core region, CdSe, where $E_g = 1.74$ eV), m_e^* and m_h^* are the effective masses of the electrons and holes, respectively, $\frac{\hbar^2}{8m_e^* d^2}$ and $\frac{\hbar^2}{8m_h^* d^2}$ are the lowest energy levels for electrons and holes, respectively, and $\frac{e^2}{8\pi\varepsilon_0 d_{eh}}$ is the Coulombic potential term with d_{eh}

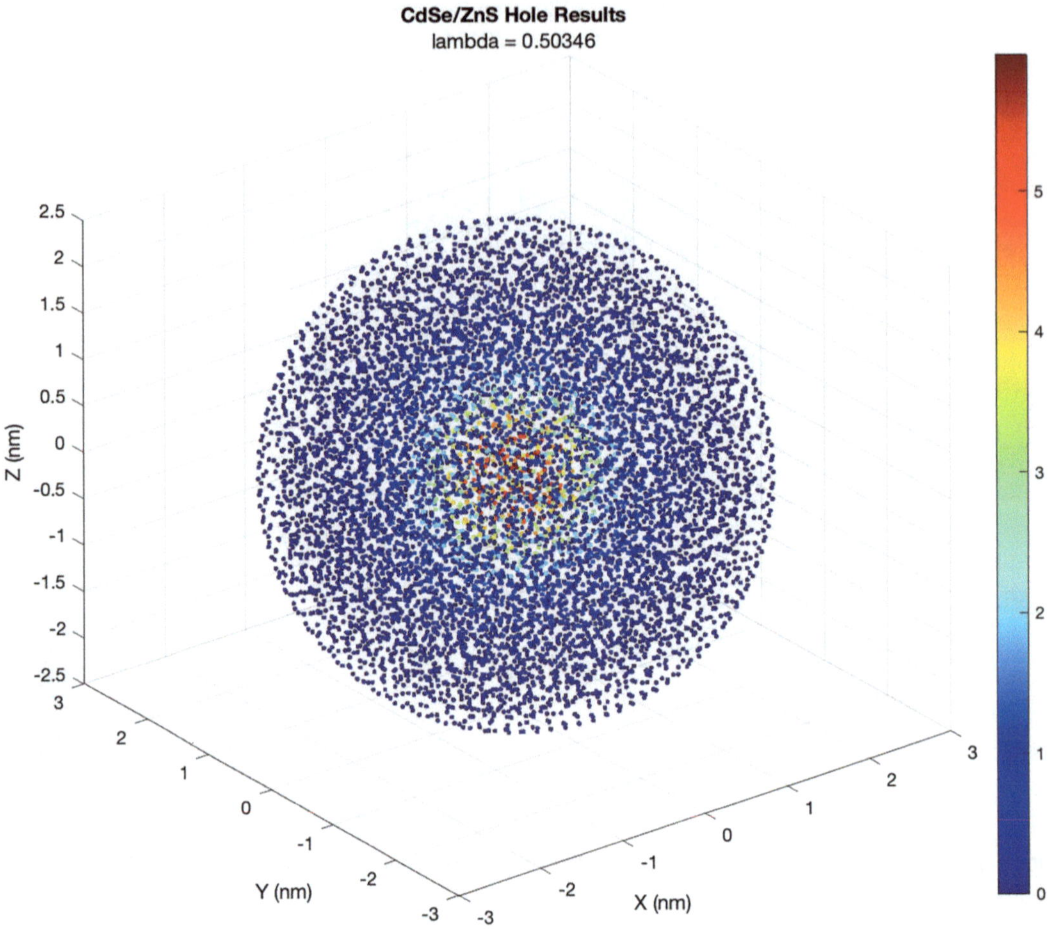

Fig. 6 Hole probability density for CdSe-ZnS. For this first energy eigenvalue, λ is 0.50346 eV and the radius is 2.5 nm.

denoting the distance wherein the charge carriers (electrons and holes) are confined. Remarkably, Eq. (4) applied to the simulation results gives $\Delta E \approx 2.28$ eV that translates to an emission wavelength $\lambda_{em} \approx 543.8$ nm. This is in excellent agreement with the manufacturer's value of 540 ± 10 nm and our measured spectroscopic results of 550 ± 30 nm emission peak for hydrophobic QD alone and 539 ± 30 nm emission peak for hydrophilic QD alone.

4 Notes

1. This helped to allow for the adjustment of the distance of the optical fiber cable to the colloidal sample to provide maximum light collection. The UV LED and the optical fiber cable were

aligned orthogonal to each other so that the collection of light would not be saturated with UV light.

2. Faxitron CellRad is relatively state-of-the art and easy to use among benchtop cell irradiators. The use of X-ray units is preferable to Gamma irradiators because Gamma irradiators are more expensive and come with radioactive sources, which require more monitoring, training, and security, than X-ray generators.
3. SOD enzymatically partitions the superoxide (O^{2-}) radical into either ordinary molecular oxygen (O_2) or hydrogen peroxide (H_2O_2). It is useful for checking molecular mechanism of the fluorescence intensity modulation and to confirm presence of ROS.
4. Catalase is an enzyme that catalyzes the reduction of hydrogen peroxide to water and oxygen.
5. Please count cells before every experiment using a standard hemocytometer. Keep cells growing at a minimum density of 200,000 cells/mL and maximum density of 1.5 million cells/mL. Keep track of cell growth. Only rapidly growing cells in the logarithmic phase of growth should be used for these experiments. We consistently centrifuged and resuspended these cells at a logarithmic phase in fresh medium at a density of 1 million cells/mL, for all our experiments. This is because the metabolic rate (and hence, ROS production) depends on cell density and on cell cycle.
6. T98G cells grow very rapidly with a population doubling time of 28 h. An important caveat is that they also adhere very rapidly to substrate. Thus, in view of *see* **Note 5**, even after resuspending the detached cells and counting out 1 million cells/mL, spectroscopic experiments must be conducted within 30 min of detachment otherwise the T98G cells reattach again and changing vessels, for instance, pipetting from flask into cuvette may be plagued with low cell density leading to results not being reproducible.
7. Our hydrogen peroxide was 50% in concentration from the manufacturer. Hence, (1.04/520)*50% = 0.1%. Adjust accordingly if starting out with hydrogen peroxide of a different concentration. Also, the 0.16 μmol/L concentration of QDs in 520 μL of deionized water, is basically unaltered by the addition of 1.04 μL of H_2O_2.
8. It is straightforward to mix cells that grow in suspension with QDs in the growth medium. However, adherent cells need to be detached into solution for spectroscopy. Decant off the old growth medium (EMEM + 10% FBS) and rinse with Dulbecco's Phosphate Buffered Saline, DPB. Add 0.5 mL Trypsin-EDTA solution to lift cells from flask, ensuring that the T25 flask is

covered by the Trypsin-EDTA solution. Incubate the flask at 37°C for 5 min. Add 5 mL of medium to deactivate Trypsin-EDTA solution. Centrifuge at 180×*g* for 5 min. Remove the supernatant and add fresh medium.

9. As expected, there is degeneracy in the eigenvalues for all $n > 1$ states, due to the chosen 3D geometry. The energy eigenvalues for $n = 2$, $n = 3$, $n = 4$ are basically the same, owing to degeneracy.

Acknowledgments

The authors acknowledge funding from the Creighton University College of Arts and Sciences (startup grant to AEE) and from a Creighton University Clare Boothe Luce Scholarship (to JA).

References

1. Alivisatos AP (1996) Semiconductor clusters, nanocrystals, and quantum dots. Science 271:933–937
2. Perini G, Palmieri V, Ciasca G et al (2020) Graphene quantum dots' surface chemistry modulates the sensitivity of glioblastoma cells to chemotherapeutics. Int J Mol Sci 21:6301. https://doi.org/10.3390/ijms21176301
3. Perini G, Palmieri V, Ciasca G et al (2021) Functionalized graphene quantum dots modulate malignancy of glioblastoma multiforme by downregulating neurospheres formation. C (Basel) 7:4. https://doi.org/10.3390/c7010004
4. Qi L, Pan T, Ou L et al (2021) Biocompatible nucleus-targeted graphene quantum dots for selective killing of cancer cells via DNA damage. Commun Biol 4:1–12. https://doi.org/10.1038/s42003-021-01713-1
5. Baker DR, Kamat PV (2010) Tuning the emission of CdSe quantum dots by controlled trap enhancement. Langmuir 26:11272–11276. https://doi.org/10.1021/la100580g
6. Lee BH (2018) Opto-electronic modulation of quantum dots by cells and bioactive molecules. Master's Thesis, Creighton University
7. Peng L, He M, Chen B, Qiao Y, Hu B (2015) Metallomics study of CdSe/ZnS quantum dots in HepG2 Cells. ACS Nano 9:10324–10334. https://doi.org/10.1021/acsnano.5b04365
8. Wang C, Gao X, Su X (2010) *In vitro* and *in vivo* imaging with quantum dots. Anal Bioanal Chem 397:1397–1415. https://doi.org/10.1007/s00216-010-3481-6
9. Goo S, Choi YJ, Lee Y, Lee S, Chung HW (2013) Selective effects of curcumin on CdSe/ZnS quantum-dot-induced phototoxicity using UVA irradiation in normal human lymphocytes and leukemia cells. Toxicol Res 9:35–42
10. Fontes A, Santos BS, (eds) (2014) Quantum dots: applications in Biology, Methods in molecular biology, vol 1197. Springer, New York, NY
11. Reczek CR, Chandel NS (2017) The two faces of reactive oxygen species in cancer. Annu Rev Cancer Biol 1:79–98
12. Sonveaux P (2017) ROS and radiotherapy: more we care. Oncotarget 2:35482–335483
13. Roy J, Galano JM, Durand T et al (2017) Physiological role of reactive oxygen species as promoters of natural defenses. FASEB J 9:3729–3745
14. Lushchak VI (2014) Free radicals, reactive oxygen species, oxidative stress and its classification. Chem Biol Interact 224:164–175
15. Sena LA, Chandel NS (2012) Physiological roles of mitochondrial reactive oxygen species. Mol Cell 2:158–167
16. Wang H-S (2016) Development of fluorescent and luminescent probes for reactive oxygen species. TrAC Trends Anal Chem 85:181–202
17. Adegoke O, Forbes PBC (2015) Challenges and advances in quantum dot fluorescent probes to detect reactive oxygen and nitrogen species: a review. Anal Chim Acta 862:1–13
18. Rosenthal SJ, Chang JC, Kovtun O, McBride JR, Tomlinson ID (2011) Biocompatible quantum dots for biological applications. Chem Biol 18:10–24
19. Algar WR, Khachatrian A, Melinger JS et al (2017) Concurrent modulation of quantum dot photoluminescence using a combination of charge transfer and Förster resonance energy transfer: competitive quenching and

multiplexed biosensing modality. J Am Chem Soc 139:363–372
20. Zrazhevskiy P, Gao X (2009) Multifunctional quantum dots for personalized medicine. Nano Today 4:414–428
21. Manshian BB, Martens TF, Kantner K et al (2017) The role of intracellular trafficking of CdSe/ZnS QDs on their consequent toxicity profile. J Nanobiotechnology 15:45
22. Lee BH, Suresh S, Ekpenyong A (2018) Fluorescence intensity modulation of CdSe/ZnS quantum dots assesses ROS during chemotherapy and radiotherapy for cancer cells. J Biophotonics 2:e201800172
23. Zhao M-X, Zhu B-J (2016) The research and applications of quantum dots as nano-carriers for targeted drug delivery and cancer therapy. Nanoscale Res Lett 11:207
24. Wojcik T, Buczek E, Majzner K et al (2015) Comparative endothelial profiling of doxoru-bicin and daunorubicin in cultured endothelial cells. Toxicol In Vitro 9:512–521
25. Cai X, Luo Y, Zhang W, Du D, Lin Y (2016) pH-Sensitive ZnO quantum dots-doxoru-bicin nanoparticles for lung cancer targeted drug delivery. ACS Appl Mater Interfaces 8:22442–22450
26. Brehwens K (2014) In vitro and in vivo aspects of intrinsic radiosensitivity. Doctoral Dissertation. Stockholm University, Stockholm
27. Vyas D, Laput G, Vyas A (2014) Chemotherapy-enhanced inflammation may lead to the failure of therapy and metastasis. Onco Targets Ther 7:1015–1023
28. Prathivadhi-Bhayankaram SV, Ning J, Mimlitz M et al (2016) Chemotherapy impedes in vitro microcirculation and promotes migration of leukemic cells with impact on metastasis. Biochem Biophys Res Commun 479:841–846

Chapter 15

Quantum Dot-Based Fluorescence-Linked Immunosorbent Assays for the Detection of Biomarkers

Yanbing Lv, Ruili Wu and Lin Song Li

Abstract

A quantum dot-based fluorescence-linked immunosorbent assay (QD-FLISA) is developed from the traditional enzyme-linked immunosorbent assay (ELISA). In QD-FLISA, quantum dots (QDs) substitute the enzyme, acting as a fluorescent label. The biosensor is built on QD-biomolecule conjugates as it signals to realize quantification through the changes in the photoluminescence (PL) of the QDs. The QD-FLISA platform exhibits high sensitivity, accuracy, and high-throughput analysis capabilities for the qualitative or quantitative detection of specific protein molecules, small molecule toxins, bacteria, or viruses.

Keywords Quantum dot, Fluorescence probe, Fluorescence-linked immunosorbent assay, Biomarker

1 Introduction

The massive demand for point-of-care (POC) diagnostics and high-throughput screening has driven the development of diagnostic technologies for obtaining more biological information from single samples with low concentration and volume [1]. Optical biosensors based on semiconductor quantum dots (QDs) are capable of meeting the aforementioned requirements due to their unique optical properties, which are well suited for high-sensitivity multiple detection. These properties include high photoluminescence (PL) intensity and quantum yields (QY), good biocompatibility, photochemical stability, and single wavelength excitation with multiple wavelength emission [2–4]. Since 1998, the improvement of QDs preparation technology, especially the successful development of biocompatible aqueous QDs, has enabled a multitude of applications in in vitro diagnostics (IVD) [5–7].

As for the traditional enzyme-linked immunosorbent assay (ELISA), the enzyme is susceptible to external conditions, highly sensitive to matrix components, and the color development of the enzyme requires a specific time frame, and these complicated operational steps

Goreti Pereira et al. (eds.), *Quantum Dots*, Methods in Molecular Biology, vol. 3025,
https://doi.org/10.1007/978-1-0716-5210-7_15,

may limit its application [8]. QDs exhibit excellent optical stability, which enables them to emit fluorescence by external light sources, saving complex operation, and color development time. As early as 2002, Goldman's group prepared QD-coupled antibodies by electrostatic adsorption of an affinity protein-antibody bridge onto the surface of CdSe/ZnS core-shell QDs, thereby successfully achieving the detection of protein toxins [staphylococcal enterotoxin B (SEB) or cholera toxin (CT)] [9]. Subsequently, quantum dot-based fluorescence-linked immunosorbent assay (QD-FLISA) was first coined in 2011 by Trapiella-Alfonso et al. They coupled the aqueous CdSe/ZnS QDs with progesterone-BSA by chemical coupling procedure with ethyl-3-(dimethylaminopropyl) carbodiimide (EDC), obtaining high-quality fluorescent probes. The quantitative detection of progesterone in milk was achieved by the competitive immunoassay method on the microplate [10]. Subsequently, Luo et al. reported a low-cost QD-FLISA based commercially QDs labeling for quantifying C-reactive protein (CRP) level by reading out the fluorescence intensity of sandwich inmmunocomplex with a microplate reader [11]. Currently, the QD-FLISA platform has been widely used for the accurate quantitative analysis of the targets, such as specific protein molecules, small molecule toxins, bacteria, or viruses [12].

Based on our synthesized hydrophobic QDs with high PL and QY, the high quality QDs or QDs nanobeads with good biocompatibility were prepared and labeled to antibodies, the QD-FLISA platform has been largely developed to realize the single or multiple quantitative detection of inflammation biomarkers and toxins [8, 13–16]. These biosensors were designed to detect quantitatively the targets based on the acquisition of fluorescence signals from QDs. The resulting ultrahigh sensitivity quantitative analysis and detection of CRP antigen were achieved through the utilization of a double-antibody sandwich method, and the results obtained were found to be in good inconsistency compared with clinical data [8]. Subsequently, the self-assembly method was used to load aqueous QDs onto the surface of silica microspheres, thereby preparing high-quality fluorescent QD microspheres. These microspheres exhibited a large surface area and could bind more antibodies, while also demonstrating enhanced sensitivity for CRP detection [14]. Highly sensitive and accurate detection of aflatoxin B1 (AFB1) were achieved by constructing a competitive FLISA using QD microspheres prepared by microemulsion method [15]. Furthermore, a FLISA platform based on red, yellow, and green QDs was constructed, and the same excitation source (450 nm) was used to excite QDs with different fluorescence peaks, thereby enabling the combined quantitative detection of three infection biomarkers within the same microplate well [16]. Considering the toxicity of Cd-based QDs, we also developed a FLISA platform based on Cd-free QDs (such as InP, CuInZnS) as fluorescent probes, which enabled accurate and quantitative detection of

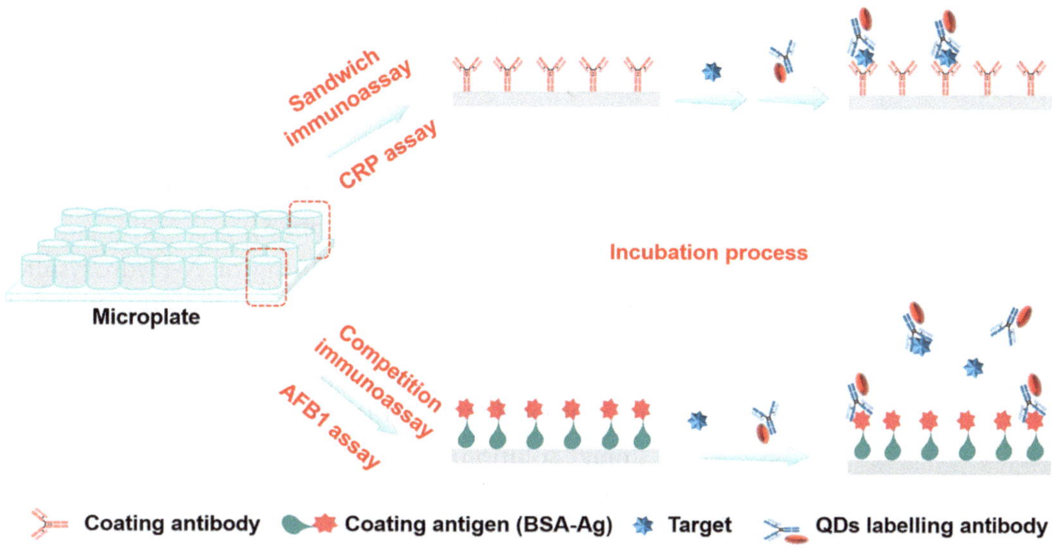

Fig. 1 Schematic diagram of QD-FLISA platforms for CRP and AFB1 detection. (Reproduced from [12] with permission from the Royal Society of Chemistry, Copyright 2023)

CRP and heart type fatty acid binding protein (H-FABP), respectively [17–19]. In view of the development trend of IVD is highly sensitive and rapidly multiple detection, the multicolor QD microspheres and fluorescent-magnetic bifunctional probes are further developed to meet the need to achieve POC detection [20].

Here, we describe the operation protocols of general QD-FLISA based common aqueous QDs (QDs@SiO_2). The CRP and AFB1 antibodies were coupled based on high-quality aqueous QDs, which served as fluorescent probes for the QD-FLISA platform. As shown in Fig. 1, the sandwich QD-FLISA and competitive QD-FLISA platforms were constructed for the quantitative detection of CRP and AFB1.

2 Materials

Prepare all solutions using ultrapure water with a resistivity of 18 MΩ·cm at 25°C, and use reagents of analytical grade.

2.1 Preparation of Aqueous QDs (QDs@SiO_2)

1. Hydrophobic QD suspension: disperse CdSe/ZnS core-shell QDs (10 mg/mL) in cyclohexane (*see* **Note 1**).
2. Ammonia solution: mix 1:1 pure ammonia (28%) with ultrapure water.
3. Carboxy reaction solution: dissolve 15 mg of succinic anhydride (SA) into 150 μL of *N*,*N*-dimethylformamide (DMF).
4. QD storage suspension: add 40 μL of pure ammonia to 3.96 mL ultrapure water.

2.2 Preparation of QD-Coupled Antibodies

1. Sodium borate (BS) buffer solution: dissolve 0.0309 g of boric acid powder in 100 mL of ultrapure water to prepare 5 mmol/L BS solution, and then adjust the pH value of the solution to 7.2, 8.0, and 9.0, respectively, with NaOH solution (1 M) using pH meter (*see* **Note 2**). Store at 4°C.
2. EDC solution: dissolve 0.434 mg of EDC in 50 μL of BS solution (5 mmol/L, pH = 7.2) (*see* **Note 3**).
3. Sulfo-NHS solution: dissolve 1.225 mg of *N*-Hydroxysulfosuccinimide (Sulfo-NHS) in 50 μL of BS solution (5 mmol/L, pH = 7.2) (*see* **Note 4**).
4. Probe blocking solution: add 10% hydrolyzed casein in BS solution (5 mmol/L, pH = 8.0). Store at 4°C.
5. Probe storage solution: add 0.1% bovine serum albumin (BSA) and 10% proclin-300 in BS solution (5 mmol/L, pH = 8.0). Store at 4°C.

2.3 Biofunctionalization of the Microplates

1. Carbonate-bicarbonate (CB) buffer solution: dissolve 0.1595 g of Na_2CO_3 and 0.293 g of $NaHCO_3$ in 100 mL ultrapure water to prepare 0.05 M CB solution, and then adjust the pH value of the solution to 9.6 with HCl solution using pH meter (*see* **Note 2**). Store at 4°C.
2. Phosphate buffered saline (PBS) solution: dissolve 0.1 g of KH_2PO_4, 1.445 g of Na_2HPO_4, 4.0 g of NaCl, and 0.1 g of KCl in 500 mL ultrapure water to prepare 0.01 M PBS solution, and then adjust the pH value of the solution to 7.4 with HCl solution using pH meter (*see* **Note 2**). Store at 4°C.
3. Blocking solution: add 0.5% BSA, 0.5% hydrolyzed casein, and 5% sucrose in PBS solution (0.01 M, pH = 7.4). Store at 4°C.
4. Wash solution: add 0.05% Tween-20 in PBS solution (0.01 mol/L, pH = 7.4). Store at 4°C.

2.4 QD-FLISA Assay

1. BS buffer solution: dissolve 123.6 mg of boric acid powder in 100 mL of water to prepare 0.02 mol/L BS solution, and then adjust the pH value of the solution to 8.0 with NaOH solution using pH meter (*see* **Note 2**). Store at 4°C.
2. CRP antigen dilution: add 10% calf serum in PBS solution (0.01 mol/L, pH = 7.4). Store at 4°C.
3. CRP antibody dilution: add 10% calf serum in PBS solution (0.01 mol/L, pH = 7.4). Store at 4°C.
4. AFB1 antigen dilution: add 10% methanol in PBS solution (0.01 mol/L, pH = 7.4). Store at 4°C.

5. AFB1 antibody dilution: add 1% BSA in BS solution (0.02 mol/L, pH = 8.0). Store at 4°C.
6. Wash solution: add 0.05% Tween-20 in PBS solution. Store at 4°C.

2.5 Instrumentation and Characterization

1. Transmission electron microscope (TEM) characterization: investigate the size and morphology of the hydrophobic QD and aqueous QD using a TEM.
2. Hydrodynamic size and zeta potential characterization: evaluate the hydrodynamic size and zeta potential of aqueous QD and QD-antibodies probes using dynamic light scattering measurements.

3 Methods

3.1 Preparation of Aqueous QDs (QDs@ SiO_2)

1. Dissolve 1.32 g of IGEPAL CO-520 in 8 mL cyclohexane, then add it into a 50 mL flask and vigorously stir until the solution remains clear. Add 2 mL of hydrophobic QD suspension (its TEM image is shown in Fig. 2a) and stir it for 30 min. Afterward, add 450 μL of ammonia solution and 160 μL of ethyl orthosilicate (TEOS) and stir it for 24 h.
2. After the reaction is finished, transfer the reaction solution to a centrifuge tube, and 2 mL of ethanol to break the microemulsion. Centrifuge the samples at 741 ×*g* for 3 min, and discard

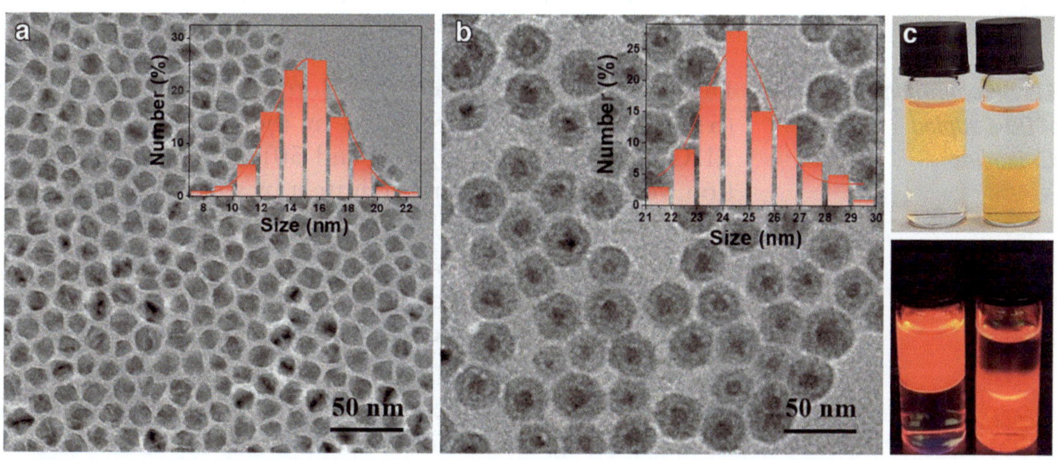

Fig. 2 TEM images of the hydrophobic QDs in n-hexane (**a**) and corresponding QDs@SiO_2 in water (**b**). Inset: size distribution data. Photographs of hydrophobic QDs (left) and corresponding aqueous QDs@SiO_2 (right) under room light (top photograph) and 365 nm ultraviolet light (bottom photograph) (upper phase is n-hexane; bottom phase is H_2O) (**c**). The sizes of hydrophobic QDs and corresponding QDs@SiO_2 are 14.3 nm and 24.8 nm, respectively.

the supernatant. Then, add 2 mL of ethanol to dissolve the precipitate by ultrasonication. Following the subsequent addition of 4 mL *n*-hexane, centrifuge the samples at 741 ×*g* for 3 min and discard the supernatant.

3. Dissolve the precipitate with 10 mL of ethanol and transfer it to a 25 mL flask. Subsequently, add 100 μL of 3-aminopropyl-triethoxysilane (APTES) and 80 μL of ammonia solution and stir it for 24 h. Then dissolve the precipitate with 10 mL of ethanol and transfer it to a 25 mL flask.
4. After the reaction is finished, take 2 mL of reaction solution in a centrifuge tube, then add 4 mL of n-hexane and centrifuge it at 741 ×*g* for 3 min, and discard the supernatant.
5. Dissolve the precipitate in 10 mL of chloroform solution and then add 150 μL of carboxy reaction solution to the flask and stir it for 24 h.
6. After the reaction is finished, add the reaction solution to a centrifuge tube and centrifuge it at 741 ×*g* for 3 min, and discard the supernatant. Dissolve the precipitate in 5 mL of ethanol solution, centrifuge at 741 ×*g* for 15 min, and discard the supernatant.
7. Blown dry the precipitate, add 4 mL of QD storage suspension and sonicate it for 30 min. Subsequently, centrifuge it in an ultracentrifuge at 2918 ×*g* for 30 min. Discard the supernatant, and add 4 mL of QD storage suspension again. The TEM image of the aqueous QDs (QDs@SiO_2) is shown in Fig. 2b. The visual evidence in Fig. 2c shows that the hydrophobic QDs were completely transferred from the organic phase to the aqueous phase, and the aqueous QDs@SiO_2 suspension was clear and transparent.

3.2 Preparation of QD-Coupled Antibodies

1. Dissolve 75 μL of aqueous QDs into 825 μL of BS solution (5 mmol/L, pH=7.2), followed by the addition of 50 μL of EDC solution (8.68 mg/mL) and 50 μL of sulfo-NHS solution, vortex and mix it, and activate it by sonication for 10 min at 4°C (*see* **Note 5**).
2. Separate the aforementioned solutions by centrifugation using a low-temperature ultracentrifuge at 2918 ×*g* for 30 min at 4°C, and discard the supernatant.
3. Dissolve the precipitate in 400 μL of BS solution (5 mmol/L, pH=8), followed by the addition of 10 μL of CRP or AFB1 antibody (5 mg/mL), and then incubate it with a rotary incubator at 4°C for overnight reaction.
4. Add 25 μL of probe blocking solution and subsequently incubate it on a rotator at 37°C for 30 min.

5. Add 6 μL of ethanolamine and subsequently incubate it on a rotator at 37°C for 30 min.
6. Expand the above solution to 1 mL with BS solution (5 mmol/L, pH=8), and then centrifuge it at 2918 ×*g* for 30 min at 4°C in a low-temperature ultracentrifuge, and discard the supernatant.
7. Expand the precipitate to 1 mL with BS solution (5 mmol/L, pH=9), centrifuge it at 2918 ×*g* for 30 min at 4°C in a low-temperature ultracentrifuge, and discard the supernatant.
8. Dissolve the precipitate in 75 μL of probe storage solution and store it in the refrigerator at 4°C.
9. The QD-antibody probes still maintain high PL intensity in Figs. 3a and d after coupling of CRP and AFB1 antibodies, their hydrodynamic size (Figs. 3b and e) and zeta potential (Figs. 3c and f) are increased, indicating the success of antibody coupling.

3.3 Construction of QD-FLISA for CRP Detection

1. Dilute the original CRP coating antibody (2.2 mg/mL) to 5 μg/mL with CB solution, add 100 μL above solution to each microplate well, and incubate in the refrigerator at 4°C for 24 h.

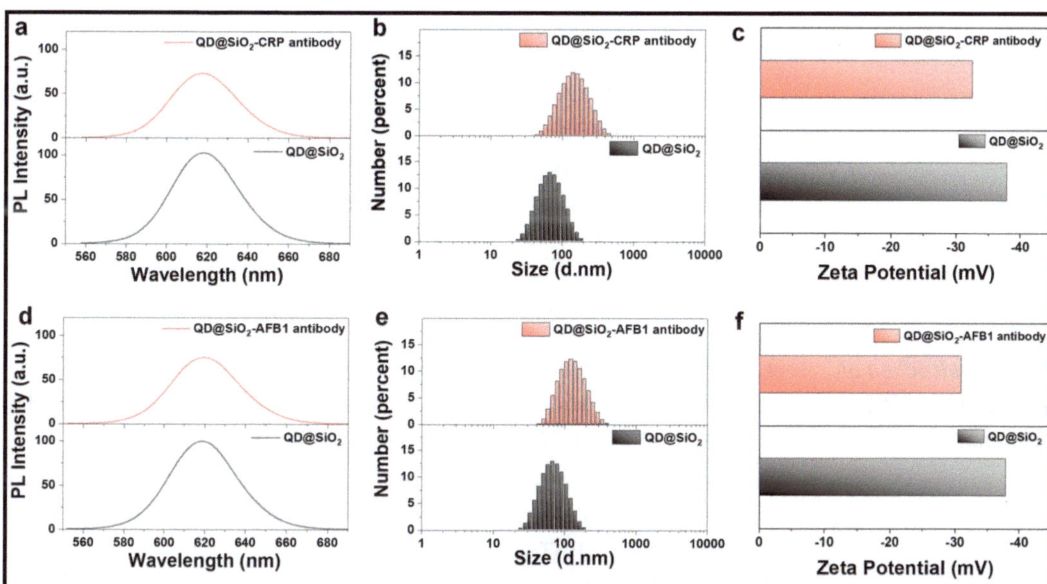

Fig. 3 PL spectra (**a**), dynamic light scattering (**b**), and zeta potential (**c**) of the aqueous QD@SiO_2 and QD@SiO_2-CRP antibody probe. PL spectra (**d**), dynamic light scattering (**e**), and zeta potential (**f**) of the aqueous QD@SiO_2 and QD@SiO_2-AFB1 antibody probe (Reproduced from [15] with permission from the ELSEVIER, Copyright 2024)

2. Remove the coating solution, wash three times with washing solution (200 μL per well at 2 min intervals), and pat dry on absorbent paper.
3. Subsequently, block each well by the addition of 110 μL of blocking solution, and then incubate it in the refrigerator at 4°C for 24 h.
4. Subsequently, shake off the sealing solution and pat dry the microplate on absorbent paper. Then, dry the microplate in a constant temperature and humidity for 1 day, and then place it in a sealed bag and store it in a 4°C refrigerator (*see* **Note 6**).
5. Remove the microplate from the refrigerator and allow it to return to room temperature (25°C) (*see* **Note 7**).
6. Dilute the CRP standard antigen (10 μg/mL) with CRP antigen dilution to the target concentration (0, 1, 5, 10, 50, 100, 500, and 1000 ng/mL), and add 100 μL/well to a biofunctionalized microplate and incubate it in a constant temperature and humidity incubator at 37°C for 30 min. Shake off the reaction solution, followed by five washes with washing solution and pat it dry on absorbent paper.
7. Dilute the QD-CRP probe with CRP probe diluent at 1:100 times, and add 100 μL/well to the corresponding microplate and incubate it in a constant temperature and humidity incubator at 37°C for 30 min. Shake off the reaction solution, followed by five washes with washing solution and pat it dry on absorbent paper.
8. Analyze the data using a fluorescence reader with the excitation at 450 nm, and measure the fluorescence spectrum. Take the maximum value at the fluorescence peak (around 620 nm).
9. The fluorescence intensity of CRP increases with the increase of its concentration, and a linear regression of fluorescence intensity (Y) and CRP antigen (X) should be performed, as shown in the Figs. 4a and b, which indicates that CRP can be quantitatively detected based on the QD-FLISA platform.

3.4 Construction of QD-FLISA for AFB1 Detection

1. Dilute the original BSA-AFB1 antigen (5 mg/mL) to 5 μg/mL with CB solution, add 100 μL above solution to each microplate well, and incubate it in the refrigerator at 4°C for 24 h.
2. Remove the coating solution, wash it three times with washing solution (200 μL per well at 2 min intervals), and pat it dry on absorbent paper.
3. Subsequently, block each well by the addition of 110 μL of blocking solution, and then incubate it in the refrigerator at 4°C for 24 h.

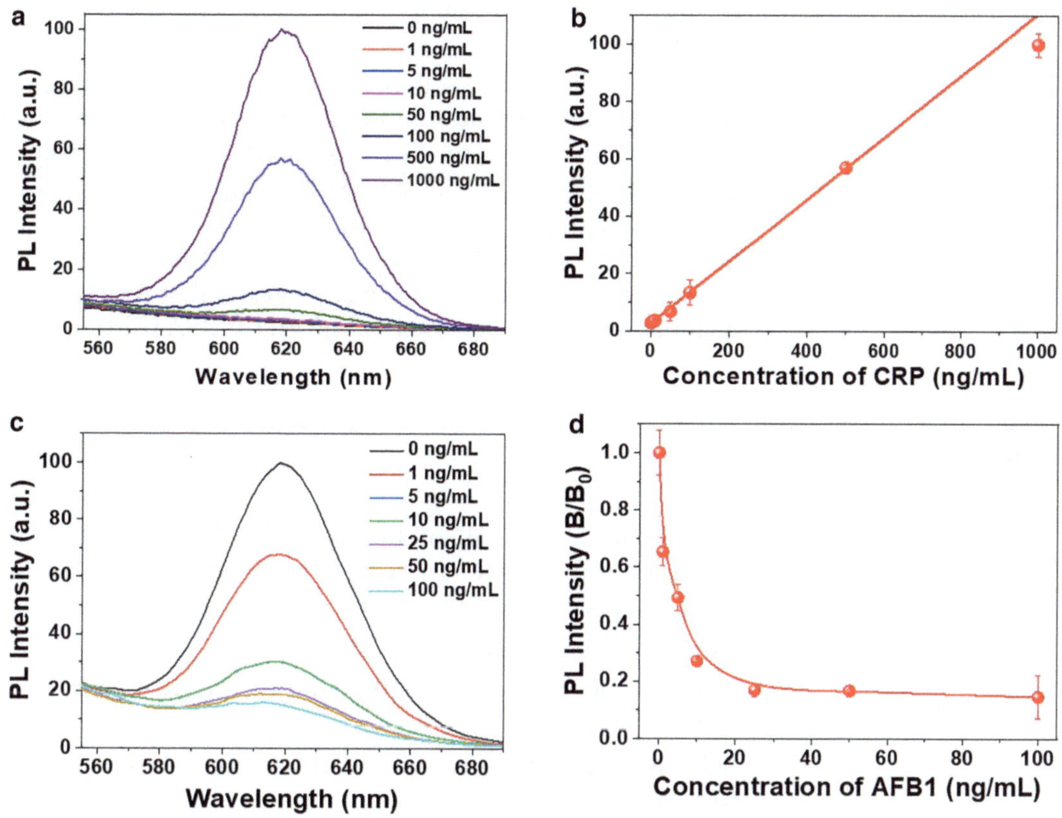

Fig. 4 PL spectra (**a**) and calibration curve (**b**) of CRP antigen by sandwich QD-FLISA platform ($n=3$). The standard curve for CRP detection was $Y=810\,X+20622$ and the linear correlation (R^2) was 0.998. PL spectra (**c**) and calibration curve (**d**) of AFB1 antigen by competitive QD-FLISA platform ($n=3$). The standard curve for AFB1 detection was $Y=(0.99-0.06)/[1+(X/2.53\text{\^{}}0.71)]+0.98$ and the linear correlation (R^2) was 0.986 (Reproduced from [15] with permission from the ELSEVIER, Copyright 2024).

4. Subsequently, shake off the sealing solution and pat dry the microplate on absorbent paper. Dry the microplate in a constant temperature and humidity for 1 day, and then place it in a sealed bag and store it in a 4°C refrigerator (*see* **Note 6**).
5. Remove the microplate from the refrigerator and allow it to return to room temperature (*see* **Note 7**).
6. Dilute the AFB1 standard antigen (2 μg/mL) with AFB1 antigen dilution to the target concentration (0, 1, 5, 10, 25, 50, and 100 ng/mL), and then add 50 μL/well to a biofunctionalized microplate. Subsequently, dilute the QD-AFB1 probe with AFB1 probe diluent at 1:100 times, and add 50 μL/well to the corresponding microplate. Incubate the microplate in a constant temperature and humidity incubator at 37°C for 30 min. Subsequently, shake off the reaction solution, followed

by five washes with washing solution and pat it dry on absorbent paper.

7. Analyze the data using a fluorescence reader with the excitation at 450 nm, and measure the fluorescence spectrum. Take the maximum value at the fluorescence peak (around 620 nm).
8. The fluorescence intensity of AFB1 decreases with the increase of its concentration, and a linear regression of fluorescence intensity (Y) and AFB1 antigen (X) should be obtained, as shown in the Figs. 4c and d, which indicates that AFB1 can be quantitatively detected based on the QD-FLISA platform.

4 Notes

1. The fluorescence peak of hydrophobic CdSe/ZnS QDs is about 620 nm, and the surface ligands can be oleic acid, oleamine, and trioctylphosphine oxide (TOPO).
2. The pH meter is calibrated at 25°C.
3. EDC powder is stored at −20°C and each time using a fresh EDC solution.
4. Sulfo-NHS powder is stored at −20°C and each time using a fresh sulfo-NHS solution.
5. The activation temperature should be selected within the range of 2–4°C, and the operation time of the activation process is not too long.
6. It is recommended that biofunctionalized microplates be stored in a sealed container to prevent moisture contamination when they are stored for an extended period of time.
7. It is essential to allow the biofunctionalized microplates to return to room temperature prior to utilization.

Acknowledgments

The authors gratefully acknowledge the financial support from the research project of the National Natural Science Foundation of China (grant numbers 22405071, 81902158, and 51972098) and Science and Technology Department of Henan Province (grant numbers 242102231045).

References

1. Jin Z, Hildebrandt N (2012) Semiconductor quantum dots for in vitro diagnostics and cellular imaging. Trends Biotechnol 30:394–403
2. Jing L, Kershaw SV, Li Y, Huang X, Li Y, Rogach AL, Gao M (2016) Aqueous based semiconductor nanocrystals. Chem Rev 116:10623–10730
3. Kagan CR, Lifshitz E, Sargent EH, Talapin DV (2016) Building devices from colloidal quantum dots. Science 353:5523
4. Arquer FPG, Talapin DV, Klimov VI, Arakawa Y, Bayer M, Sargent EH (2021) Semiconductor quantum dots: technological progress and future challenges. Science 373:8541
5. Chan WCW, Nie SM (1998) Quantum dot bioconjugates for ultrasensitive nonisotopic detection. Science 281:2016–2018
6. Chi X, Huang D, Zhao Z, Zhou Z, Yin Z, Gao J (2012) Nanoprobes for in vitro diagnostics of cancer and infectious diseases. Biomaterials 33:189–206
7. Xu G, Zeng S, Zhang B, Swihart MT, Yong K, Prasad PN (2016) New generation cadmium-free quantum dots for biophotonics and nanomedicine. Chem Rev 116:12234–12327
8. Lv Y, Wu R, Feng K, Li J, Mao Q, Yuan H, Shen H, Chai X, Li LS (2017) Highly sensitive and accurate detection of C-reactive protein by CdSe/ZnS quantum dot-based fluorescence-linked immunosorbent assay. J Nanobiotechnol 15:35
9. Goldman ER, Balighian ED, Mattoussi H, Kuno MK, Mauro JM, Tran PT, Anderson GP (2002) Avidin: a natural bridge for quantum dot-antibody conjugates. J Am Chem Soc 124:6378–6382
10. Trapiella-Alfonso L, Costa-Fernández JM, Pereiro R, Sanz-Medel A (2011) Development of a quantum dot-based fluorescent immunoassay for progesterone determination in bovine milk. Biosens Bioelectron 26:4753–4759
11. Luo Y, Zhang B, Chen M, Jiang T, Zhou D, Huang J, Fu W (2012) Sensitive and rapid quantification of C-reactive protein using quantum dot-labeled microplate immunoassay. J Transl Med 10:24
12. Lv Y, Fan J, Zhao M, Wu R, Li LS (2023) Recent advances in quantum dot-based fluorescence-linked immunosorbent assays. Nanoscale 15:5560–5578
13. Lv Y, Li J, Wu R, Wang G, Wu M, Shen H, Li LS (2018) Silica-encapsulated quantum dots for highly efficient and stable fluorescence immunoassay of C-reactive protein. Biochem Eng J 137:344–351
14. Li J, Lv Y, Li N, Wu R, Xing M, Shen H, Li LS, Chen X (2019) Robust synthesis of bright multiple quantum dot-embedded nanobeads and its application to quantitative immunoassay. Chem Eng J 361:499–507
15. Xiao W, Ye Y, Li N, Xu H, Lv Y, Wu R, Shen H, Li LS (2024) Preparation of 3D pomegranate-like quantum dot nanobeads and its application to highly sensitive detection of aflatoxin B1. Microchem J 200:110380
16. Lv Y, Li N, Wang L, Fan J, Xing H, Shi Y, Yu S, Wu R, Shen H, Li LS (2024) Tailored three-color quantum dots nanobeads for multiplexed detection with tunable detection range and multilevel sensitivity of signal-amplified immunosensor. Talanta 269:125416
17. Xu Y, Lv Y, Wu R, Li J, Shen H, Yang H, Zhang H, Li LS (2021) Sensitive immunoassay based on biocompatible and robust silica-coated Cd-free InP-based quantum dots. Inorg Chem 60:6503–6513
18. Jin Q, Zhang X, Zhang L, Li J, Lv Y, Li N, Wang L, Wu R, Li LS (2023) Fabrication of CuInZnS/ZnS quantum dot microbeads by a two-step approach of emulsification-solvent evaporation and surfactant substitution and its application for quantitative detection. Inorg Chem 62:3474–3484
19. Zhang L, Xu H, Zhang X, Chen X, Lv Y, Zhang R, Wang L, Wu R, Shen H, Li LS (2024) Highly sensitive, stable InP quantum dot fluorescent probes for quantitative immunoassay through nanostructure tailoring and biotin-streptavidin coupling. Inorg Chem 63:4604–4613
20. Fan J, Shi Y, Lv Y, Li N, Wu R, Shen H, Li LS (2025) Tailored controllable bifunctional fluorescent-magnetic nanobeads for new generation fast-response immunosensor with dual-signal amplification. Sens Actuat B Chem 423:136804

Chapter 16

Glutathione-Coated Quantum Dots as Immobilized Substrates for the Evaluation of Glutathione-Dependent Enzymes: Glutathione Reductase and Glutathione *S*-Transferase

Duygu Aydemir and Nuriye Nuray Ulusu

Abstract

Quantum dots (QDs), owing to their unique optical properties and tunable surface functionalities, have emerged as powerful tools in biomedical research, particularly in diagnostics, drug delivery, and bioimaging. Among them, glutathione (GSH)-coated silver sulfide (Ag_2S) QDs represent a promising class of biocompatible nanomaterials with near-infrared (NIR) fluorescence, making them highly suitable for biological applications. This chapter presents a novel approach utilizing GSH-coated Ag_2S QDs as immobilized substrates and substrate analogues for the evaluation of glutathione-dependent enzymes glutathione reductase (GR) and glutathione *S*-transferase (GST). Through kinetic analyses, both purified enzymes and biologically relevant enzyme sources, such as rat liver lysate and cultured cell lysates, were employed to investigate the interaction efficiency and enzymatic activity with QD-bound GSH. The results demonstrated that GSH-coated QDs not only participate in enzyme-catalyzed redox cycling but also retain functionality comparable to free GSH or oxidized glutathione (GSSG), particularly under physiological conditions. Additionally, microscopic imaging confirmed the internalization of these QDs in mammalian cells, indicating their potential for in vivo enzymatic assays and theranostic applications. Overall, this study introduces a robust and innovative nanobiotechnological platform to explore enzyme–substrate interactions and redox regulation, with significant implications for biosensing, cancer research, and nanomedicine.

Keywords Semiconductor nanocrystals, Immobilized enzyme substrates, Enzyme activity, Fluorescence, Nanomedicine, Bio-imaging

1 Introduction

Quantum dots (QDs) are emerging nanomaterials that exhibit unique electronic, optical, and biocompatible properties under physiological conditions, facilitating various biomedical applications [1–4]. They are extensively used in fluorescence resonance energy transfer (FRET), gene technology, medical imaging, targeting therapies, cellular tracking, and bio-detection. These applications also encompass

Goreti Pereira et al. (eds.), *Quantum Dots*, Methods in Molecular Biology, vol. 3025,
https://doi.org/10.1007/978-1-0716-5210-7_16,

biosensing, drug delivery, live imaging, tumor imaging, tissue engineering, and cancer therapies, underscoring their significance in diagnostics, preventive oncology, and theragnostic [1–4]. The use of QDs in the detection and treatment of various coronaviruses holds significant promise due to their low toxicity and exceptional surface properties, which distinguish them from other nanomaterials [5]. For instance, when combined with fluorescent probes, QDs emerge as highly effective semiconductor nanomaterials for monitoring cellular processes and identifying viral infections. Their unique characteristics make them a valuable tool in advancing diagnostic and therapeutic strategies for viral diseases. For instance, there are limited studies on using QDs to diagnose SARS-CoV-2, but functionalized QDs could play a significant role in the future safe nano-therapies for viral infections [6].

QDs can be functionalized with ligands to enhance their interaction with the immune system. This functionality enables QDs to target specific immunoglobulin E (IgE) and effector cell receptors, facilitating a precise immune response [3]. The surface of QDs can be modified with various molecules, including hydrophilic, hydrophobic, and amphiphilic substances. These functionalized surfaces can then be linked to drugs, DNA, proteins, antibodies, tripeptides, for instance, glutathione, and other compounds for targeted applications [3, 7, 8]. Investigating enzyme activities using various substrates, substrate analogs, and immobilized substrates is essential for understanding enzymatic mechanisms and specificity [9, 10]. Using immobilized substrates and enzymes provide further insights into enzyme–substrate interactions and the potential for selective inhibition, modulation, and biomedical applications [11]. This approach facilitates the explanation of enzyme–substrate interactions and contributes to developing novel therapeutic strategies [12].

Immobilization of substrates onto nanomaterials, such as QD, may have several advantages in enzymatic and biosensing applications; for instance, immobilized substrates may enhance the interaction efficiency between the enzyme and substrate [13]. Due to their high specific surface area and surface free energy, nanoparticles enable the immobilization of biomolecules without causing denaturation or loss of biological activity, primarily through covalent bonding and electrostatic interactions [14], allowing them to interact with different structures within living organisms [15]. For instance, QDs enable stable immobilization of thiolated aptamers via sulfur–sulfur affinity, exploiting the interaction between thiol groups (–SH) of aptamers and the metal ions in QDs. The study concludes that QDs are effective immobilized substrates for aptamer-based sensors due to their chemical affinity, stability, and signal enhancement capabilities, particularly in conjunction with electroactive probes for sensitive and selective small-molecule detection [16].

Glutathione-coated silver sulfide quantum dots (GSH-Ag_2S QDs) emerge as promising nanomaterials for biomedical applications, particularly in the field of cancer diagnostics and therapeutics. Due to their surface functionalization with GSH, these QDs are not only biocompatible but also intrinsically fluorescent in the near-infrared (NIR) region, making them suitable for deep-tissue imaging. Furthermore, their GSH shell provides a model to investigate glutathione-dependent enzymatic processes and redox regulation. GSH is a tripeptide composed of glutamine, cysteine, and glycine, and is recognized as a major intracellular antioxidant that plays a crucial role in maintaining cellular redox homeostasis via regulating detoxification of xenobiotics, antioxidant defense, cell proliferation, and cell death pathways. Through its thiol group, GSH readily conjugates with electrophilic compounds and detoxifies reactive oxygen species (ROS), heavy metals, and other harmful substances, protecting cellular components from oxidative damage [5, 12]. The intracellular concentration of GSH and the GSH/GSSG ratio are also critically important in the pathophysiology of cancer. Elevated levels of GSH have been reported in various malignancies, including breast, ovarian, head and neck, and lung cancers, potentially contributing to tumor cell survival, proliferation, and resistance to chemotherapy. Conversely, decreased GSH levels have been observed in brain and liver tumors, indicating a tumor-type-specific regulation of glutathione metabolism [17].

This chapter describes the methodology employed in utilizing GSH-coated Ag_2S QDs as immobilized substrates or substrate analogues for GSH-dependent enzyme activity for glutathione reductase (GR) and glutathione *S*-transferase (GST) for the first time in the literature. Commercially available GR and GST enzymes are employed to investigate the substrate properties of GSH-coated QDs through kinetic analyses. Additionally, rat liver lysate is utilized as a biological source of GR and GST to further assess the substrate-analog potential of GSH-coated QDs in a more physiologically relevant context for in vivo systems. Given that the thiol group of GSH binds covalently to the Ag_2S surface, the conjugated form of GSH participates in GR-catalyzed and GST-catalyzed redox cycling. This investigation not only provides insight into the interaction between GR/GST and surface-immobilized GSH but also offers a platform for studying diseases associated with impaired glutathione metabolism.

2 Materials

All solutions should be prepared using ultrapure water, obtained by further purification of deionized water to achieve a resistivity of 18 MΩ cm at 25°C. Only analytical-grade reagents should be used. Substrates, enzymes, and detection kits should be stored at 4°C or

−20°C. Cell lysates treated with QDs should be stored at −80°C until experiments are carried out, and all experimental steps should be performed on ice to avoid protein degradation. Proper adherence to institutional and governmental waste disposal regulations is essential when handling and discarding chemical waste.

2.1 Cell Culture

1. Cell lines: human breast adenocarcinoma (MCF7), human cervical carcinoma (HeLa), ductal pancreatic adenocarcinoma (CFPAC-1), and human embryonic kidney (293T).
2. 96-well tissue culture plates.
3. Culture medium DMEM (for 293T cells) and RPMI-1640 (for MCF7, HeLa, and CFPAC-1 cells) supplemented with 10% fetal bovine serum (FBS) and 1% penicillin-streptomycin (pen-strep) solution.
4. MTT solution (5 mg/mL): add 5 mg of (3-(4,5-dimethylthiazol-2-yl)-2,5-diphenyltetrazolium bromide) in 1 mL of 1 M phosphate-buffered saline (PBS), vortex thoroughly.
5. Mix analytical-grade dimethyl sulfoxide (DMSO) and ethanol (EtOH) 1:1 v:v.

2.2 Evaluation of Cell Morphology and Internalization of QDs via Microscopic Techniques

1. Culture medium DMEM (for 293T cells) and RPMI-1640 (for MCF7, HeLa, and CFPAC-1 cells) supplemented with 10% FBS and 1% pen-strep solution.
2. 3 mL glass-bottom Petri dishes (suitable for high-resolution microscopy).
3. 4% Paraformaldehyde prepared in 1 M PBS (for fixation).
4. DAPI (4′,6-diamidino-2-phenylindole) solution is used as a nuclear stain at a final concentration of 2 μg/mL, prepared in a 1:1 (v/v) mixture of glycerol and PBS.
5. Inverted confocal fluorescence microscope.
6. Oil immersion objective lens (63×).
7. Fluorescence filter sets for QDs: tailored to specific QD excitation/emission spectra (e.g., 510–550 nm excitation, ≥710 nm emission for Ag_2S QDs). Fluorescence filter sets for DAPI: excitation 325–375 nm, emission 435–485 nm.

2.3 Cell Lysate Preparation for GR and GST Enzyme Source

1. 25 cm^2 polystyrene canted-neck tissue culture flasks.
2. Complete growth medium RPMI (HeLa, CFPAC-1) and DMEM (293T) (with 10% FBS and 1% pen-strep, unless otherwise specified).
3. 1 M PBS, pH 7.4, ice-cold (precooled on ice for 1 h).
4. 200 mM sodium phosphate (Na_2PO_4) buffer, pH 7.4, ice-cold (precooled on ice for 1 h).
5. Commercial protease inhibitor cocktail tablets—added freshly to Na_2PO_4 buffer to prevent protein degradation.

6. Sterile cell scrapers.
7. 1.5 mL microcentrifuge tubes, prechilled in the 4°C refrigerator overnight.
8. Probe sonicator—for lysate homogenization (10-s pulse, on ice).

2.4 Preparation of Tissue Lysate for the GR and GST Enzyme Source

1. Wistar albino rat (*Rattus norvegicus*), 24–28 weeks old, 350–450 g.
2. Glass cages with stainless steel covers.
3. Air-conditioned animal room (12-h light/dark cycle, 21 ± 4°C, 50% ± 5% humidity).
4. Standard pellet diet.
5. 200 mM sodium phosphate (Na_2PO_4) buffer, pH 7.4, ice-cold (precooled on ice for 1 h).
6. Commercial protease inhibitor cocktail tablets—added freshly to Na_2PO_4 buffer to prevent protein degradation.
7. IKA Ultra-Turrax homogenizer.
8. Refrigerated centrifuge capable of 105,000 ×*g* (set at 4°C).

2.5 Measurement of Protein Concentration

1. Pierce™ BCA protein assay kit.
2. Bovine serum albumin (BSA) standard (included in the kit or purchased separately).
3. Distilled or deionized water.
4. Cell lysate samples (CFPAC-1, 293 T, HeLa cells).
5. Liver homogenate samples.
6. 96-well microplate.
7. Microplate reader or spectrophotometer (capable of reading at 562 nm).
8. Incubator or water bath (set to 37°C or as recommended by protocol).

2.6 Kinetic Studies to Evaluate the Substrate Analog Potential of the GSH-Coated Quantum Dots via GR Activity

1. NADPH solution (0.2 mM) in ultrapure water.
2. GSSG solution (0.5, 0.75, 1, 1.5, and 2 mM) in ultrapure water.

2.7 Kinetic Studies to Evaluate the Substrate Analog Potential of the GSH-Coated Quantum Dots via GST Activity

1. Sodium phosphate buffer (Na_2PO_4, 100 mM) at pH 6.5.
2. Purified GST enzyme (2 μg/mL).
3. 1-Chloro-2,4-dinitrobenzene (CDNB, 1 mM).
4. GSH and GSH-functionalized QDs (0 to 2 mM).
5. GST (0.4–2 μg/mL).

3 Methods

3.1 Evaluation of Cell Viability Via 3-(4,5-Dimethylthiazol-2-Yl)-2,5-Diphenyltetrazolium Bromide (MTT) Assay

After synthesizing and characterizing the GSH-coated Ag_2S QDs of interest, their cytotoxicity is evaluated using the MTT assay.

1. Cultured the cells at densities ranging from 12,500 (CFPAC-1 and MCF7) to 25,000 (HeLa and 293T) cells per well in 96-well plates, depending on the cell type, in culture medium supplemented with 10% FBS and 1% pen-strep.
2. Incubate the cells for 24 h at 37°C in a humidified atmosphere with 5% CO_2 for attachment and recovery (*see* **Note 1**).
3. After incubation, aspirate the culture medium and replace it with fresh medium containing varying concentrations of QDs, ranging from 5 to 600 μg/mL based on their determined content.
4. Incubate the cells for 24 h under the same conditions.
5. Following nanoparticle treatment, add 20 μL of the MTT solution (5 mg/mL) to each well, and incubate the plates for 4 h to allow viable cells to reduce MTT to formazan crystals via mitochondrial dehydrogenase activity.
6. After incubation, remove the medium containing excess MTT carefully, and add 200 μL of a 1:1 (v/v) mixture of DMSO/EtOH to each well to solubilize the formazan crystals.
7. Measure the absorbance at 600 nm with a reference correction at 630 nm in a microplate reader (*see* **Note 2**).
8. Calculate the cell viability as a percentage by normalizing the absorbance of treated wells (mean of six replicates per condition) to that of untreated control wells (cells exposed only to medium). One-way analysis of variance (ANOVA), followed by Tukey's multiple comparisons test, is used to determine statistical significance among treatment groups. A p-value less than 0.05 is considered statistically significant.

3.2 Evaluation of Cell Morphology and Internalization of QDs via Microscopic Techniques

Morphological assessment of the cellular response to QD exposure is conducted using routine light microscopy. Cells are visually examined every 24 h for morphological alterations, including shape, size, adherence, and membrane integrity changes. Fluorescence microscopy is employed to evaluate intracellular uptake and the imaging potential of the applied QDs (Fig. 1).

1. Seed the cells under investigation into 3 mL glass-bottom Petri dishes—suitable for high-resolution imaging—at a density of 175,000 cells per dish.

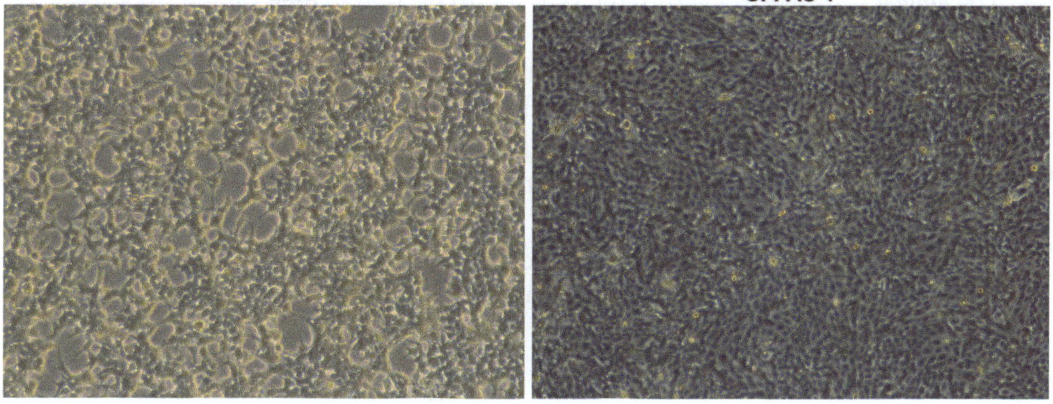

Fig. 1 Light microscopy images of the 293T and CFPAC-1 cells

2. After a 24-h incubation period to allow for adhesion and spreading, treat the cells with GSH-coated Ag_2S QDs (e.g., Ag_2S, CdSe, ZnS, InP other QDs with GSH coating can be used and tested as well) at a final 100 μg/mL concentration.
3. Incubate the QDs with the cells for 24 h to facilitate internalization.
4. Following incubation, wash the cells three times with sterile PBS to remove any uninternalized QDs.
5. Fix the cells with 4% paraformaldehyde in PBS for 20 min at room temperature (~25°C).
6. After fixation, stain the nuclei with DAPI solution (2 μg/mL) for 10 min, serving as a nuclear counterstain.
7. Subsequent wash with PBS to remove excess dye and maintain the samples in 3 mL of PBS to prevent drying before imaging.
8. Acquire fluorescence images using an inverted confocal fluorescence microscope with oil immersion at 63× magnification (Fig. 1). Two distinct filter sets are used to simultaneously visualize QDs and stained nuclei: one tailored to the excitation and emission spectra of the QDs, typically in the near-infrared (NIR) region (e.g., excitation $\lambda = 510$–550 nm, emission $\lambda \geq$ 710 nm for Ag_2S QDs), and one for DAPI (excitation $\lambda = 325$–375 nm, emission $\lambda = 435$–485 nm) (*see* **Note 3**).
9. After acquiring the fluorescence images for both channels, process them using ImageJ software. Channels are color-coded, merged, and contrast-enhanced as needed to assess co-localization of QDs within cellular compartments and evaluate their suitability for bioimaging applications (Fig. 2). Untreated control cells (maintained under identical conditions but without QD exposure) are included to distinguish QD-specific fluorescence from background or cellular autofluorescence [18].

3.3 Preparation of Cell Lysates as a Source of GR and GST Enzymes for Evaluating the Substrate Capability of GSH-Coated Quantum Dots

1. Culture the cells in 25 cm^2 polystyrene canted-neck tissue culture flasks under standard conditions until they reach approximately 80%–90% confluency.
2. After cells reach the desired confluency, place the flasks immediately on ice to halt cellular activity.
3. Wash the cells twice with ice-cold PBS to remove residual medium.
4. Subsequently, add 500 μL of ice-cold Na_2PO_4 buffer (pH 7.4) to each flask Na_2PO_4 buffer (pH 7.4) to each flask,

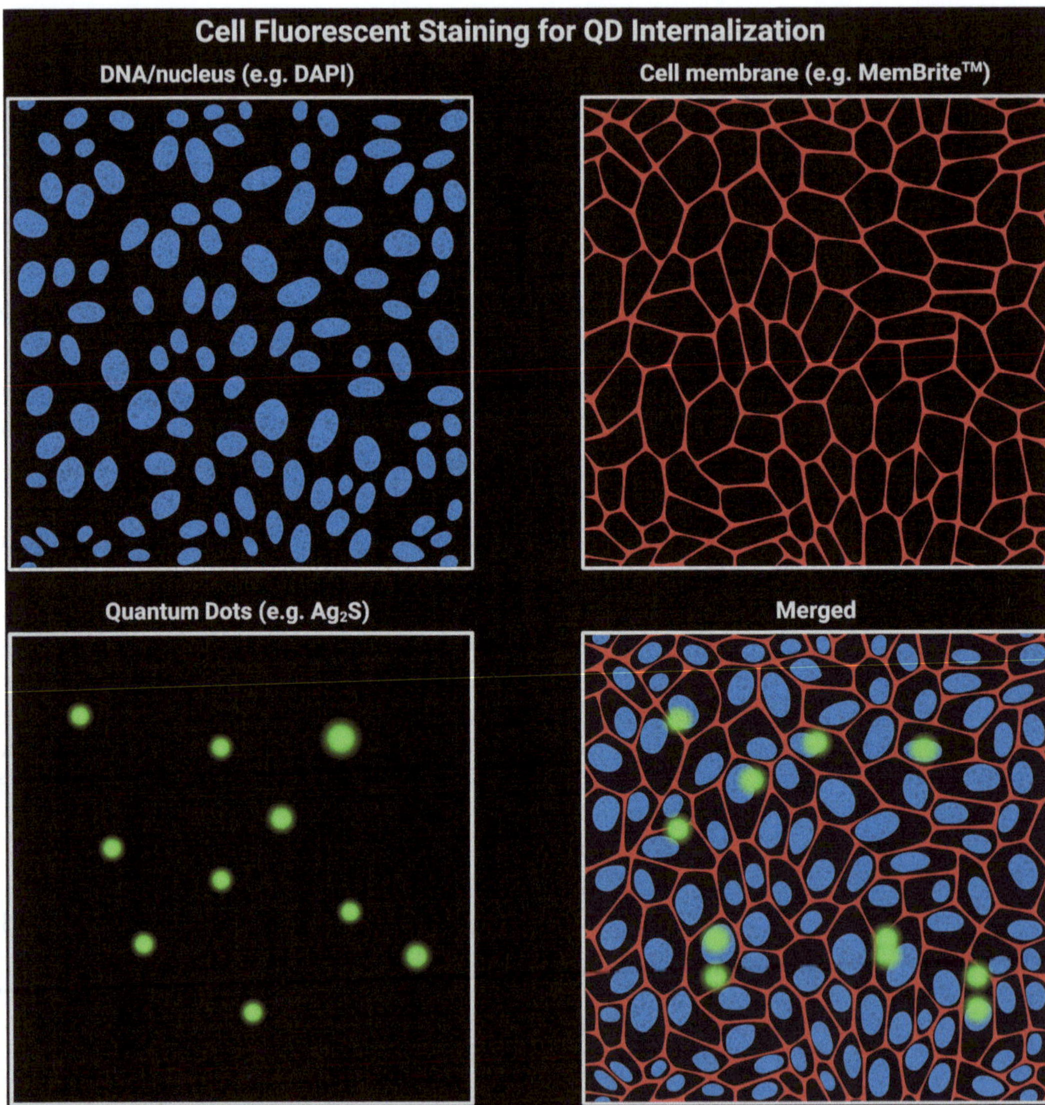

Fig. 2 Evaluation of the quantum dot (QD) internalization via microscopy following cell fluorescent staining

supplemented with a protease inhibitor cocktail to prevent protein degradation.

5. Lyze the adherent cells in situ and gently scrape off the surface using a sterile cell scraper (*see* **Note 4**).
6. Transfer the resulting cell suspensions to prechilled 1.5 mL microcentrifuge tubes.
7. Sonicate the samples on ice for 10 s using a probe sonicator to ensure complete lysis and homogenization. Centrifuge the lysates at 105,000 ×*g* for 1 h at 4°C to separate the soluble protein fraction from cellular debris.
8. Collect the supernatants containing the clarified cell lysates carefully and store them at −80°C until further biochemical or proteomic analyses are performed.

3.4 Preparation of Tissue Lysate for the GR and GST Enzyme Source

1. A male Wistar albino rat (*Rattus norvegicus*), aged between 24 and 28 weeks and weighing 350–450 g, is obtained from the Experimental Animals Production Center.
2. All animals are housed in glass cages with stainless steel lids, maintained in an air-conditioned room under a 12-h light/dark cycle, with a temperature of 21 ± 4°C and relative humidity of 50% ± 5%.
3. Throughout the experimental period, rats had ad libitum access to standard pellet feed and tap water (*see* **Note 5**).
4. Animals are sacrificed under anesthesia via guillotine, and liver tissues are collected.
5. Wash liver tissue three times with ice-cold 1 M PBS buffer to eliminate blood residues.
6. Weigh and homogenize tissue samples using an IKA Ultra-Turrax homogenizer in a 1:3 (w/v) ratio of ice-cold 100 mM Na_2PO_4 buffer (pH 7.4) supplemented with a protease inhibitor cocktail.
7. The homogenates are centrifuged at 105,000 ×*g* for 1 h at 4°C. The resulting supernatant is collected and used as the source of GST and GR enzymes.

3.5 Measurement of Total Soluble Protein Concentration in Cell Lysates and Tissue Samples

1. The total protein concentration of cell lysates (CFPAC-1, HeLa, 293 T) and rat liver homogenates is quantified using the bicinchoninic acid (BCA) assay (following the manufacturer's protocol).
2. BSA is used as the standard to generate a calibration curve for protein quantification.
3. Color change in the samples and standards is measured spectrophotometrically at 562 nm.

3.6 Kinetic Studies to Evaluate the Substrate Analog Potential of the GSH-Coated Quantum Dots via GR Activity

1. The enzymatic reaction is initiated by adding the GR enzyme (prepared in Sect. 3.3) and monitored by measuring the decrease in absorbance at 340 nm using a UV-visible spectrophotometer at 37°C for 30 s. All reactions are performed in duplicate, and a linear increase in absorbance is observed throughout the time course of the assay.
2. GR activity is assessed at a fixed GR enzyme concentration of 0.5 μg/mL and a saturating concentration of NADPH (0.2 mM) using various substrate concentrations (0.5, 0.75, 1, 1.5, and 2 mM) of GSSG, GSH-Ag_2S, and GSH in separate reaction setups (*see* **Notes 6 and 7**).
3. The effect of varying NADPH concentrations (0.05, 0.1, 0.2, 0.3, and 0.4 mM) on GR activity is evaluated at a fixed substrate concentration of 1 mM (GSH-Ag_2S, GSH, and GSSG in separate reactions) and an enzyme concentration of 0.5 μg/mL.
4. The impact of increasing GR enzyme concentrations (0.5, 0.75, 1, 1.5, and 2 μg/mL) on substrate activity is examined at saturating NADPH concentrations of 0.2 mM and 1 mM substrate concentration (GSH-Ag_2S, GSH, and GSSG in separate reactions).
5. 10 μL of rat liver lysate (protein concentration: 24 mg/mL) and 100 μL of cell lysate (protein concentration: 1.3 mg/mL) are tested at saturating NADPH concentration (0.2 mM), and different concentrations (0.5, 0.75, 1, 1.5, 2 mM) of GSSG, GSH-Ag_2S, and GSH are studied in separate reactions (*see* **Notes 6 and 7**).

3.7 Kinetic Studies to Evaluate the Substrate Analog Potential of the GSH-Coated Quantum Dots via GST Activity

1. The assay is performed in a reaction mixture containing 100 mM sodium phosphate buffer (Na_2PO_4, pH 6.5), 2 μg/mL of purified GST enzyme, CDNB (1 mM), and GSH or GSH-functionalized QDs as substrates at final concentrations ranging from 0 to 2 mM. The total reaction volume is maintained consistently across all samples.
2. The enzymatic reaction is initiated by adding GST and monitored by measuring the increase in absorbance at 340 nm using a UV-visible spectrophotometer at 37°C for 30 s. All reactions are performed duplicate, and a linear increase in absorbance is observed throughout the time course of the assay [19].
3. To investigate the effect of enzyme concentration on reaction kinetics, increasing concentrations of GST (0.4–2 μg/mL) are tested at fixed optimal substrate concentrations of CDNB (1 mM) and GSH or GSH-coated QDs (1 mM). Absorbance at 340 nm is recorded to monitor the enzymatic activity. Each reaction is carried out in duplicate, and the response progress

remains linear throughout the measurement period. GST activity is further evaluated using cell lysates.

4. Assays are conducted at an optimal CDNB concentration of 1 mM, using either GSH or GSH-coated QDs as substrates at final concentrations of 0.25, 0.5, 0.75, and 1 mM. For each assay, 10 μL of rat liver lysate (protein concentration: 24 mg/mL) and 100 μL of cell lysate (protein concentration: 1.3 mg/mL) are used as sources of GST isozymes. Reactions are performed in duplicates under standard assay conditions, and absorbance changes at 340 nm are used to quantify enzyme activity.

4 Notes

1. Cells are cultured under standard conditions (37°C, 5% CO_2, humidified atmosphere) in a complete growth medium appropriate for each cell line.
2. The intensity of the resulting purple color correlates with the number of metabolically active cells. The absorbance at 630 nm is subtracted from that at 600 nm to correct for background interference.
3. Filters are selected based on the specific photophysical properties of the QDs employed, ensuring optimal signal capture with minimal spectral overlap.
4. Cell lysate preparation should be performed on ice, and after scraping cells, a precooled centrifuge should be used to collect the cells. The protease inhibitor cocktail should be freshly added to the PBS to dissolve the cells. Following preparation of cell lysates, all experiments should be performed on ice or at 4°C to avoid protein degradation.
5. All animal procedures needs to be conducted according to the ethical standards and approved by the institution ethics.
6. QDs should not be left at room temperature for too long; if possible, use freshly synthesized QDs to treat cells. Otherwise, the coating of the QDs can be degraded or oxidized.
7. Add enzymes or cell lysates at the end to evaluate the enzymatic reaction and measure immediately to prevent losing the enzymatic activity range.

References

1. Panja A, Patra P (2023) A review on quantum dots (QDs) and their biomedical applications. 4open 6:1. https://doi.org/10.1051/fopen/2022020
2. Sarkar S, Srivastava T, Sahoo O et al. (2024) Applications of quantum dots in preventive oncology. Asian Pac J Cancer Prev 25:747–756. https://doi.org/10.31557/APJCP.2024.25.3.747
3. Mohkam M, Sadraeian M, Lauto A, Gholami A, Nabavizadeh SH, Esmaeilzadeh H, Alyasin S (2023) Exploring the potential and safety of quantum dots in allergy diagnostics. Microsyst Nanoeng 9:145. https://doi.org/10.1038/s41378-023-00608-x
4. Arab K, Jafari A, Shahi F (2024) The role of graphene quantum dots in cutting-edge medical therapies. Polym Adv Technol. 35. https://doi.org/10.1002/pat.6571
5. Aydemir D, Hashemkhani M, Durmusoglu EG, Acar HY, Ulusu NN (2019) A new substrate for glutathione reductase: glutathione coated Ag_2S quantum dots. Talanta 194:501–506. https://doi.org/10.1016/j.talanta.2018.10.049
6. Rabiee N, Ahmadi S, Soufi GJ et al. (2022) Quantum dots against SARS-CoV-2: diagnostic and therapeutic potentials. J Chem Technol Biotechnol 97:1640–1654. https://doi.org/10.1002/jctb.7036
7. Parascandolo A, Rappa F, Cappello F et al. (2017) Extracellular superoxide dismutase expression in papillary thyroid cancer mesenchymal stem/stromal cells modulates cancer cell growth and migration. Sci Rep 7:41416. https://doi.org/10.1038/srep41416
8. Aydemir D, Hashemkhani M, Acar HY, Ulusu NN (2020) Evaluation of the biocompatibility of the GSH-coated Ag_2S quantum dots in vitro: a perfect example for the non-toxic optical probes. Mol Biol Rep 47:4117–4129. https://doi.org/10.1007/s11033-020-05522-3
9. Ulusu NN (2015) Curious cases of the enzymes/Neobiča istorija enzima. J Med Biochem 34:271–281. https://doi.org/10.2478/jomb-2014-0045
10. Ulusu NN (2015) Evolution of enzyme kinetic mechanisms. J Mol Evol 80:251–257. https://doi.org/10.1007/s00239-015-9681-0
11. Aydemir D, Çakır S, Özdemir N, Ulusu NN (2024) Evaluation of the antimicrobial activity of triple enzyme-embedded organic–inorganic hybrid nanoflowers (hNFs) in comparison with powerful antimicrobial agent chitosan. Curr Microbiol 81:359. https://doi.org/10.1007/s00284-024-03884-6
12. Aydemir D, Ulusu NN (2020) Comment on the: molecular mechanism of CAT and SOD activity change under MPA-CdTe quantum dots induced oxidative stress in the mouse primary hepatocytes (Spectrochim Acta A Mol Biomol Spectrosc. 2019 Sep 5; 220:117104). Spectrochim Acta A Mol Biomol Spectrosc 229:117792. https://doi.org/10.1016/j.saa.2019.117792
13. Fu Y, Liu T, Wang H, Wang Z, Hou L, Jiang J, Xu T (2024) Applications of nanomaterial technology in biosensing. J Sci Adv Mater Devices 9:100694. https://doi.org/10.1016/j.jsamd.2024.100694
14. Zhang W, Zhang Y, Lu Z, Nian B, Yang S, Hu Y (2023) Enhanced stability and catalytic performance of laccase immobilized on magnetic graphene oxide modified with ionic liquids. J Environ Manage 346:118975. https://doi.org/10.1016/j.jenvman.2023.118975
15. Bamburowicz-Klimkowska M, Poplawska M, Grudzinski IP (2019) Nanocomposites as biomolecules delivery agents in nanomedicine. J Nanobiotechnol 17:48. https://doi.org/10.1186/s12951-019-0479-x
16. Hua M, Li P, Li L, Huang L, Zhao X, Feng Y, Yang Y (2011) Quantum dots as immobilized substrate for electrochemical detection of cocaine based on conformational switching of aptamer. J Electroanal Chem 662:306–311. https://doi.org/10.1016/j.jelechem.2011.08.017
17. Hashemkhani M, Bilici K, Muti A, Sennaroglu A, Acar HY (2020) Ag_2S-glutathione quantum dots for NIR image guided photothermal therapy. New J Chem 44:5419–5427. https://doi.org/10.1039/C9NJ04608A
18. Aydemir D, Hashemkhani M, Acar HY, Ulusu NN (2020) Evaluation of the biocompatibility of the GSH-coated Ag_2S quantum dots in vitro: a perfect example for the non-toxic optical probes. Mol Biol Rep 47:4117–4129. https://doi.org/10.1007/s11033-020-05522-3
19. Aydemir D, Hashemkhani M, Acar HY, Ulusu NN (2019) In vitro interaction of glutathione S-transferase-pi enzyme with glutathione-coated silver sulfide quantum dots: a novel method for biodetection of glutathione S transferase enzyme. Chem Biol Drug Des 94:2094–2102. https://doi.org/10.1111/cbdd.13614

INDEX

Goreti Pereira et al. (eds.), *Quantum Dots*, Methods in Molecular Biology, vol. 3025,
https://doi.org/10.1007/978-1-0716-5210-7,